THE DETERMINATION OF MOLECULAR STRUCTURE

BY

P. J. WHEATLEY

SECOND EDITION

DOVER PUBLICATIONS, INC.
NEW YORK

Published in Canada by General Publishing Company,
Ltd., 30 Lesmill Road, Don Mills, Toronto, Ontario.
Published in the United Kingdom by Constable and Com-
pany, Ltd., 10 Orange Street, London WC2H 7EG.

This unabridged Dover edition, first published in 1981, is
printed by special arrangement with Oxford University Press.

Manufactured in the United States of America
Dover Publications, Inc.
180 Varick Street
New York, N.Y. 10014

Library of Congress Cataloging in Publication Data

Wheatley, P   J
    The determination of molecular structure.

    Reprint of the 2d ed. published by the Clarendon
Press, Oxford, England.
    Includes bibliographies and indexes.
    1. Molecular structure. I.   Title.
QD461.W56 1981   541.2'2   80-21988
ISBN 0-486-64068-X (pbk.)

PREFACE TO THE SECOND EDITION

THE words 'molecular structure' mean different things to different people. In this second edition, as in the first, they are taken to mean a knowledge of bond lengths and bond angles, or, if these cannot be determined, at least a knowledge of molecular symmetry. The powerful diagnostic methods such as infra-red and nuclear magnetic resonance spectroscopy, whose main use is the determination of the way in which atoms are connected by chemical bonds without reference to geometrical details, are discussed only in so far as they can provide information on 'molecular structure' as defined above.

The changes from the first edition are not substantial. Certain points have been clarified and infelicities removed, mainly as a result of suggestions from people who were kind enough to write to me about the first edition. The chapters on neutron diffraction and nuclear magnetic resonance have been modified to include advances in these rapidly changing fields.

I should like to thank Professors K. Hedberg and P. M. Harris for permission to reproduce illustrations that did not appear in the first edition.

P. J. W.

September 1966

PREFACE TO THE FIRST EDITION

THE object of any scientific experiment is the elimination of further experimentation. In physical chemistry this aim can be stated more precisely, since the ultimate object of any experiment is the elucidation of the forces within and between molecules. Once these forces are fully understood, it will be possible, provided the computations can be performed, to calculate the result of any chemical experiment. Fortunately, at least for the experimentalists, there seems to be no doubt that it will be some little time before this goal is achieved. In the meantime one promising and important road by which an understanding of the forces between atoms may be reached is the investigation of the shapes and sizes of molecules. Indeed, Professor J. E. Wertz has stated that 'Pursuit of the details of molecular structure and molecular environment is the occupation of all chemists part of the time and part of the chemists all of the time'.

This book is designed to give an introductory survey of the main physico-chemical methods that have been devised for the determination of molecular structures. I believe it to be important for a student to know which particular method is likely to be most suitable for the determination of the structure of a particular molecule, and also how much information that method can be expected to give. Furthermore, it is necessary to be able to assess the reliability of the molecular dimensions obtained. In order to satisfy these requirements the emphasis throughout has been placed on the scope and limitations of the various methods. Only short accounts are given of actual experimental techniques. Long mathematical derivations have been omitted, and basic equations are often introduced with no formality other than a bare statement of their validity. Finally, no attempt has been made to account for the results of structure determinations in terms of current theories of valency.

For students who wish to probe more deeply into any one particular method there are many excellent books and articles available. I have relied heavily on these standard works, to which reference is made in the text or at the end of the appropriate chapter.

I am grateful to the following for permission to reproduce illustrations or tables from original papers or books: Professors E. R. Andrews, E. F. Barker, S. H. Bauer, J. M. Bijvoet, Sir Lawrence Bragg, L. O. Brockway, G. Giacomello, W. Gordy, O. Hassel, G. Herzberg, H. H. Nielsen, G. E. Pake, N. F. Ramsey, M. T. Rogers, C. G. Shull, K. N. Trueblood, and J. E. Wertz; Drs. Isabella L. Karle, G. E. Bacon, C. W. Bunn, W. T. Eeles, F. C. Nachod, R. E. Richards, B. P. Stoicheff, H. W. Thompson, H. Viervoll, and L. A. Woodward; the Councils of the Chemical Society, the Faraday Society, and the Royal Society; the editors of the Canadian Journal of Physics, Physical Review, and Reviews of Modern Physics; the American Chemical Society, the American Institute of Physics, the International Union of Crystallography, and Svenska Teknologföreningen; and Messrs. Academic Press Inc., G. Bell & Sons, Ltd., Gebrüder Borntraeger, Butterworth, D. B. Centen, Cornell University Press, The Kynoch Press, D. Van Nostrand Company, Inc., John Wiley & Sons, Inc., and The Williams & Wilkins Company.

The Weissenberg and powder photographs reproduced in Chapter VI were taken by Mr. J. J. Daly and Mr. E. L. Lippert respectively. Mr. R. Clemson of the University of Leeds both sat for and took Fig. 1.3. Miss E. C. Ochsner of Monsanto Research S.A. assisted with the preparation of the manuscript.

I would like to thank Dr. J. A. S. Smith who read and made valuable comments on preliminary versions of most chapters. I am indebted to Professor E. G. Cox, F.R.S., of the University of Leeds and to Dr. N. T. Samaras of Monsanto Research S.A. for their encouragement and patience. Finally I must express my appreciation of the staff of the Clarendon Press for their help.

P. J. W.

Zürich
January 1958

CONTENTS

I

SYMMETRY

1.1. Introduction

FIG. 1 (a) shows a schematic drawing of a hydrogen molecule. Fig. 1 (b) illustrates a molecule of hydrogen chloride. Figs. 2 (a) and 2 (b) also show molecules of hydrogen and hydrogen chloride. Fig. 2 (b) clearly differs from Fig. 1 (b) in that the molecule has been turned round and, reading from left to right, we have

(a)   (b)

FIG. 1. Homonuclear and heteronuclear diatomic molecules:
(a) hydrogen, H_2, (b) hydrogen chloride, HCl.

(a)   (b)

FIG. 2. Homonuclear and heteronuclear diatomic molecules:
(a) hydrogen, H_2, (b) hydrogen chloride, HCl.

drawn ClH rather than HCl. But if we compare Fig. 2 (a) with Fig. 1 (a), we see that they are identical and we cannot tell whether the molecule has been turned round or not. To put the matter more precisely, we can say that, if we rotate the hydrogen molecule through 180° about an axis normal to the internuclear axis, we produce an orientation that is indistinguishable from the original. In the previous sentence we have introduced two important ideas. The first is that a molecule may have two (or more) orientations that are indistinguishable. If this is so, we say that the molecule possesses *symmetry*. The second idea is that in order to convert one orientation of a symmetrical molecule into an indistinguishable orientation we have to move, or imagine we move, the molecule in some way. The operation whereby an orientation is changed into an indistinguishable one is known as a *symmetry operation*. The operation described above for the hydrogen molecule involved rotation, but this is not the only type of symmetry operation. If we look at a pair of hands (Fig. 3) we see that the right hand is related to the left hand (or vice versa) in that one is the mirror image of the other. Fig. 3 possesses symmetry because one half can be produced from the other by reflexion. We say that Fig. 3 possesses a

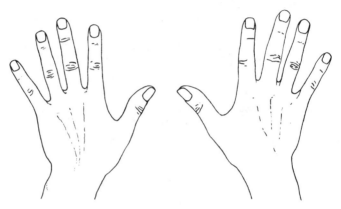

FIG. 3. A pair of hands related by a vertical mirror plane.

symmetry element, the element of symmetry being a mirror plane which lies normal to the plane of the paper and half-way between the hands. In the case of the hydrogen molecule the symmetry element that we discussed was a rotation axis. A symmetry element tells us what sort of symmetry operation must be carried out in order to relate one portion of a symmetrical object to another.

Another important aspect of symmetry is illustrated in Fig. 4. In Fig. 4 (a) we have drawn a system of atoms to represent the planar boron trifluoride molecule. This molecule has symmetry because we can turn it through 120° about an axis normal to the plane of the paper through the boron atom and obtain an indistinguishable orientation. In Fig. 4 (b) we have drawn three such molecules. Each molecule in Fig. 4 (*b*) still has, of course, its own symmetry; but the three molecules are related to each other in the same way that the three fluorine atoms in a single molecule are related. We must, therefore, extend our idea of

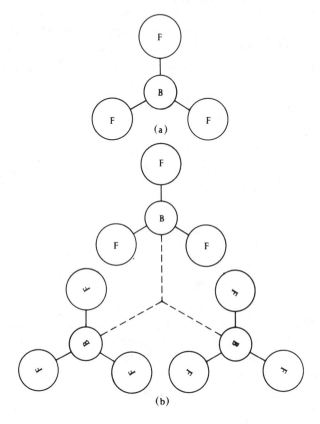

FIG. 4. The molecule of boron trifluoride, BF_3: (*a*) a single molecule, (*b*) three molecules related by symmetry in the same way that the atoms in a single molecule are related.

symmetry to cover a regular arrangement of a number of molecules as well as the arrangement of atoms within a single molecule. We shall see later that when we are working in the gas phase we are interested only in the symmetry of an isolated molecule, but when we are dealing with solids we are interested in the symmetrical arrangement of a number of molecules as well as in the symmetry of an individual molecule.

We shall find, as we proceed, that it is impossible to discuss molecular structure without mentioning symmetry. There are several reasons for this, and they will emerge later. At the moment we may just mention the practical point that it is easy to determine the dimensions of a molecule like methyl cyanide (Fig. 5), which contains six atoms and is highly sym-

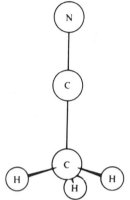

FIG. 5. The molecule of methyl cyanide, CH_3CN.

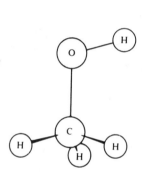

FIG. 6. The molecule of methyl alcohol, CH_3OH.

metrical, whereas it is extremely difficult to find the dimensions of a molecule like methyl alcohol (Fig. 6), which in many ways is very similar to methyl cyanide except that the alcohol has symmetry only if the hydroxyl hydrogen atom is locked in the eclipsed or staggered position, which is probably not the case. The first question that a chemist asks when he has to find the structure of a molecule is: Does it possess any symmetry? Because if it does, his job will almost always be easier than if it does not.

Since symmetry is so intimately bound up with molecular structure, it is necessary to be able to think easily and confidently in terms of symmetry. This task is simplified by the fact that symmetry is omnipresent. The external human form, if we ignore minor blemishes, possesses a plane of symmetry. This book, if we disregard the printing, has a plane of symmetry which lies between the central pages. The mirror plane cannot be seen between these pages in the same sense that the Earth's equator can never be seen, but it is there just the same. Wherever we look we see symmetry, and it can be quite a fascinating occupation to pick out the symmetry of a building or a wallpaper or the decorations on an ornament.

In order to make full use of the ideas of symmetry, we must first be familiar with the different symmetry elements. We are fortunate here in that there is only a limited number of different types. We could discuss these symmetry elements formally and mathematically in terms of group theory, and this is certainly the most elegant and probably the best way of dealing with symmetry. However, for a chemist, a more descriptive approach is probably more illuminating, and this is the one we adopt here.

1.2. Point symmetry and space symmetry

Symmetry may perhaps best be defined as regular arrangement. As we have seen, the arrangement may involve repetition of identical parts of a single object as in Figs. 1 (a), 4 (a), or 5, or it may involve the repetition of a number of identical objects as in Fig. 4 (b). Symmetry is best described in terms of the symmetry elements, which portray the different ways in which repetition may be achieved. We have to consider the two cases mentioned above. In the first, repetition of identical parts of a single object, we are interested in the symmetry about a point, in the so-called *point symmetry*. We are then, in this book, considering the symmetry of an isolated molecule. In the second case we have to consider a regular array of identical molecules, and thus we have to include the possibility of symmetry elements involving translation. We are then dealing with the symmetry of a crystal, with the so-called *space symmetry*.

In order to appreciate fully the effect of symmetry elements on an object, we need to start with an object that has no symmetry itself; we need an *asymmetric unit*. An atom is particularly bad from this point of view since it has the highest possible symmetry, spherical symmetry. Consequently we must replace our atoms by something that has no symmetry. We shall choose a coin as our asymmetric unit, but we must remember all the time that our coin may represent a single atom or a group of atoms or a whole molecule, depending on the system that is under discussion. In fact, for a reason which will become apparent, we cannot use a real coin to demonstrate symmetry elements, so we shall simplify it somewhat (Fig. 7). We shall replace the 'head' by the letter P (Fig. 7 (a)) and the 'tail' by the

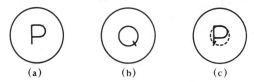

(a)   (b)   (c)

FIG. 7. A figure suitable for an asymmetric unit: (a) obverse side, (b) reverse side, (c) obverse and reverse sides.

letter Q (Fig. 7 (b)). Apart from this alteration, our coin is like a real British coin in that, if it is turned over from right to left (or left to right), the Q is right way up and right way round. If the coin were transparent, it would appear as in Fig. 7 (c), the dotted Q signifying that it lies underneath. It would be no use marking our coin with H for head and T for tail because the coin would then itself possess symmetry (Fig. 8). If this coin

FIG. 8. A similar figure which is not an asymmetric unit, since it possesses symmetry.

were divided down the middle, the left half would be the mirror image of the right. This coin would possess an internal plane of symmetry, and the action of this plane of symmetry would confuse the effects of external symmetry elements. In other words, our asymmetric unit would now not be the coin, but half the coin. We shall have to make our illustrations in terms of our coins in order to prevent ambiguities due to internal symmetry creeping in, but, wherever possible, we shall reproduce a real molecule so that the relation to chemistry shall not be lost.

1.3. Symmetry elements

For a description of the point symmetry we need to consider only four types of symmetry elements: the *centre of symmetry*, the *mirror plane, rotation axes*, and *alternating* or *inversion axes*. Unfortunately two different notations are used to describe the symmetry elements. The newer type is the Hermann–Mauguin notation, and is perhaps to be preferred since it is equally applicable to the description of point symmetry and space symmetry. Crystallographers usually employ this system. However, the older Schoenflies notation is just as informative as the Hermann–Mauguin notation for the description of point symmetry, and is still widely used by spectroscopists, who are usually interested only in isolated molecules. We shall, therefore, include the Schoenflies symbol in brackets after the Hermann–Mauguin symbol.

Centre of symmetry, $\bar{1}$ (C_i). In Fig. 9 two of our coins are related by a *centre of symmetry* X. In order to obtain one coin from the other, we join each point of one coin to the centre of symmetry and produce the line for an equal distance on the other side. In order to add a third dimension to the plane of the paper, we have marked one coin $+$ to denote that it lies above the plane of the paper. Since the action of the centre of symmetry will put the other coin below the plane of the paper, we have marked it $-$. Fig. 9 may be considered in the manner just mentioned, or the whole diagram may be regarded as one molecule, in which case each coin is half the molecule and the two halves are related by the centre of symmetry. We say that one half of the molecule is obtained from the other by *inversion* through the centre of symmetry. Although centrosymmetric molecules are quite common, there are not many molecules that possess only a centre of symmetry and no other symmetry element. A substituted ethane HXYC—CYXH in which each pair of similar atoms is in the *trans* position is a suitable example, and Fig. 10 shows *meso*-ClBrHC—CHBrCl.

Mirror plane, m (C_s). In Fig. 11 the two coins are related by a *mirror plane of symmetry* lying vertically through the line AB. In order to generate one coin (or half of the molecule) from the

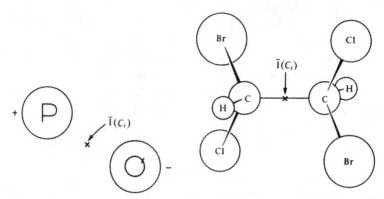

Fig. 9. A centre of symmetry.

Fig. 10. The molecule of *meso*-ClBrHC—CHBrCl.

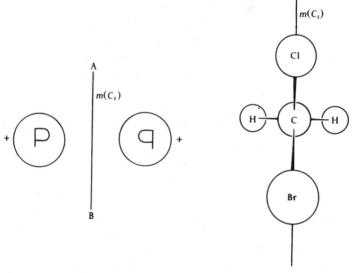

Fig. 11. A mirror plane of symmetry.

Fig. 12. The molecule of chloro-bromomethane, CH_2BrCl.

other, we drop a perpendicular from each point on to the mirror plane and produce for an equal distance on the other side. All our construction lines are, therefore, parallel, whereas in the case of the centre of symmetry they converged. If one coin lies above the plane of the paper, so does the other. Fig. 12 shows the molecule CH_2BrCl which possesses only a mirror plane of symmetry. It will be seen that the carbon, bromine, and chlorine atoms lie in the mirror plane. They are allowed to do this because atoms are themselves symmetrical. The asymmetric unit of CH_2BrCl is thus one hydrogen atom, half a carbon atom, half a bromine atom, and half a chlorine atom. Atoms which actually lie in an element of symmetry are said to be in *special positions*. We shall have more to say about special positions when we discuss crystals (section 7.3).

Rotation axes, n (C_n). *n* is known as the *order of the rotation axis*. If a molecule has a rotation axis of order *n*, rotation through $2\pi/n$ or $360/n°$ produces an orientation indistinguishable from the original. In an isolated molecule a rotation axis can have any order from 1 to ∞. If $n = 1$ we have the *identity operation* which merely involves rotation of the molecule through $2\pi = 360°$. If the only element of symmetry that a molecule possesses is the identity operation, we say that the molecule has no symmetry. If $n = 2$ we speak of a *twofold* or *dyad axis*, and rotation of the molecule through $2\pi/2 = 180°$ produces an orientation indistinguishable from the original. In Fig. 13 (*a*) the two coins are related by a twofold axis lying in the plane of the paper. If one coin lies above the plane of the paper, the other must lie below. In Fig. 14 (*a*) the coins are related by a twofold axis normal to the plane of the paper and passing through *O*. In this case both coins lie on the same side of the paper. Again there are few molecules that possess only a twofold axis, but we can use as an example the twisted form of the oxalate ion as it occurs in ammonium oxalate hemihydrate. This form of the ion has three mutually perpendicular twofold axes and is shown in Figs. 13 (*b*) and 14 (*b*) with one of the axes in orientations corresponding to Figs. 13 (*a*) and 14 (*a*). If $n = 3, 4,...$, we speak of *triad, tetrad, ... axes*, or *threefold, four-fold, ... axes*. The boron trifluoride molecule shown in Fig. 4 has a threefold axis, although it has other symmetry elements as well. The order of the rotation axis can theoretically have any value for an isolated molecule, but in fact it seldom rises above six,

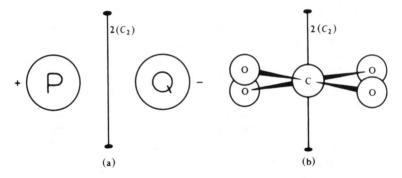

FIG. 13. A twofold axis: (a) axis lying in the plane of the paper, (b) a distorted form of the oxalate ion, $C_2O_4^{2-}$, with the twofold axis lying in the plane of the paper.

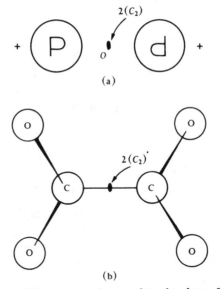

FIG. 14. A twofold axis: (a) axis normal to the plane of the paper, (b) a distorted form of the oxalate ion, $C_2O_4^{2-}$, with the twofold axis normal to the plane of the paper.

except for the case of linear molecules which have an *infinite rotation axis* coincident with the internuclear axis. The rotation axis is of infinite order because rotation of a linear molecule around the internuclear axis by any fraction of 360° produces an indistinguishable orientation. This, of course, is another consequence of the spherical symmetry of atoms.

The fourth type of symmetry element is described in rather different terms in the Schoenflies and in the Hermann–Mauguin systems. The Schoenflies system is based on *alternating* or *rotary-reflexion axes* (S_n), whereas the Hermann–Mauguin

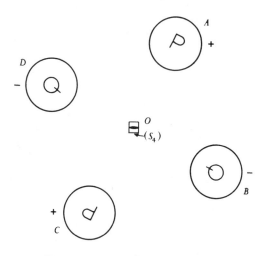

FIG. 15. A fourfold alternating axis.

system is based on *inversion* or *rotary-inversion axes* $\bar{n}$. Only when the order of the axis is divisible by four are these the same.

Alternating axes, (S_n). A fourfold alternating axis (S_4) is illustrated in Fig. 15. The axis lies normal to the plane of the paper through O. The coin at the starting-point A, lying above the plane of the paper, is rotated clockwise through $2\pi/4 = 90°$ and then reflected across the plane of the paper to give the coin at B lying below the paper. A similar operation performed on the coin at B produces one at C, and so on back to A. If the whole of Fig. 15 is regarded as one molecule, the asymmetric unit is one quarter of the molecule.

Inversion axes, $\bar{n}$. A fourfold inversion axis $\bar{4}$ is shown in Fig. 16. The coin at the starting-point A is rotated clockwise through 90° and then inverted through the centre O to give the coin at B lying below the plane of the paper. A similar operation performed on the coin at B gives the one at C. C then gives D and D gives A again. It will be seen that, except for the lettering,

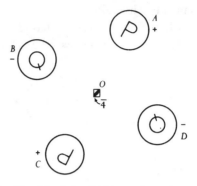

FIG. 16. A fourfold inversion axis.

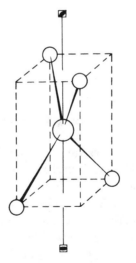

FIG. 17. Atoms related by a fourfold alternating or inversion axis.

Figs. 15 and 16 are identical, although they were generated in rather different ways. In Fig. 17 the result of a fourfold alternating or inversion axis is displayed in a different manner. The other common alternating axes have the equivalent inversion axes shown opposite:

Schoenflies	Hermann–Mauguin
S_1	$\bar{2} \equiv m(C_s)$
S_2	$\bar{1}$
S_3	$\bar{6}$
S_4	$\bar{4}$
S_6	$\bar{3}$

It should be noticed that, in one respect, pure rotation axes differ from all other types of symmetry elements. In Figs. 13 and 14 it can be seen that the operation of a rotation axis produces a coin which is superposable on the original. All other symmetry elements produce a mirror image of the original coin. This is why we cannot use real coins to illustrate the symmetry elements: real coins are all manufactured in the form of a single enantiomorph. Let us suppose that the figures represent actual molecules and let us consider the action of molecules on plane polarized light. We know that, if a molecule has no symmetry, it will rotate the plane of polarization of the light. If a molecule possesses only a pure rotation axis then the effect of the asymmetric unit on the plane of polarization will be enhanced by the remaining parts of the molecule related to the asymmetric unit by the rotation axis. On the other hand, if the molecule possesses any other sort of symmetry element, the effect of the asymmetric unit on the plane of polarization will be nullified by the effect of the mirror image of the asymmetric unit. In other words, a pure rotation axis is the only sort of symmetry element that a molecule may possess and still show optical activity. This point will be met again in Chapter X.

1.4. Combination of symmetry elements

It has probably seemed surprising that we have had to select rather obscure chemical compounds with which to illustrate the symmetry elements, but by this time the reason will no doubt be apparent. Most simple molecules will have more than one of the possible elements of symmetry. Thus the water molecule (Fig. 18) has two mirror planes at right angles, and a twofold axis where the mirror planes intersect. In fact, whenever mirror planes intersect, their line of intersection is necessarily a rotation axis, and, in general, combinations of symmetry elements produce other symmetry elements. The automatic production of the extra symmetry elements means that it is unnecessary to specify every single element in order to describe completely the symmetry of a molecule. For water all we need to say is that the molecule has two mirror planes at right angles. We have no need to mention the twofold axis because we know it must be there. We shall return to this redundancy of symmetry

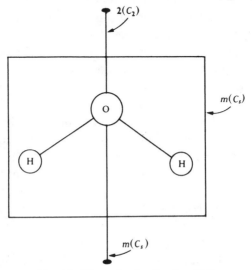

FIG. 18. The molecule of water, H_2O.

elements in the next section, but first we shall give some more examples of molecules that possess several elements of symmetry.

In order to describe the pyramidal ammonia molecule (Fig. 19), it is sufficient to say that it possesses a threefold axis and a mirror plane lying in the rotation axis. The threefold axis will automatically produce two more mirror planes, so that the full symmetry is a threefold axis, and three mirror planes whose line of intersection coincides with the threefold axis. The symmetry of the benzene molecule (Fig. 20) can be completely specified by saying that it possesses a sixfold axis, a plane of symmetry at right angles to this axis, and a plane of symmetry lying in the

rotation axis. But again this is not all the symmetry elements
that the molecule has. The presence of the above three elements

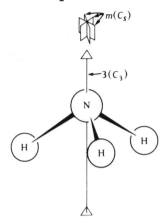

FIG. 19. The molecule of ammonia, NH₃.

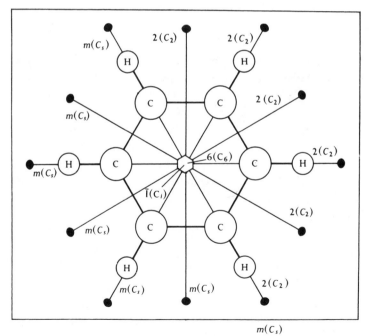

FIG. 20. The molecule of benzene, C₆H₆.

demands that altogether the molecule has a centre of symmetry, a sixfold axis, six twofold axes, and seven mirror planes. Finally a tetrahedral molecule like methane has three fourfold inversion axes, four threefold axes, and six mirror planes, although again it is not necessary to mention all of these in order to specify completely the symmetry of a tetrahedral molecule. We have

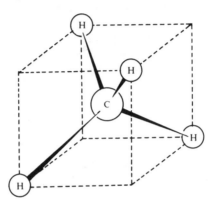

Fig. 21. The molecule of methane, CH_4.

drawn methane inside a cube in Fig. 21. With the help of Figs. 17 and 19 it should be possible to pick out the different symmetry elements. It is important to notice that a tetrahedral molecule does not possess a centre of symmetry.

It might seem that the number of ways in which symmetry elements can combine is unlimited. In fact this is not so. It can be shown mathematically (and this is where group theory is so valuable) that only certain combinations are possible. However, we must remember that, for an isolated molecule, there is an infinite number of rotation (and inversion or alternating) axes, so that there must be an infinite number of possible point symmetries. This is still rather a lot, but we shall see in the next section how this infinite number is reduced in practice to a manageable size.

1.5. Crystallographic point groups

When we come to consider a point in a crystal, a restriction is set on the possible symmetry about this point by the requirements that a crystal shall consist of regularly repeating units in three dimensions and that the environment of each unit shall be identical. In order to satisfy these requirements it can be proved that no axes of order greater than six are possible and, moreover, that a fivefold axis is excluded. This does not mean, of course, that a molecule with a fivefold axis cannot form a crystal. It means that the surrounding molecules cannot be related by a fivefold axis. If instead of an infinite number of rotation axes we admit only those with $n = 1, 2, 3, 4$, and 6, it can be shown that there are only thirty-two possible ways of combining symmetry elements, and these are known as the thirty-two *crystallographic point groups*. A list of these thirty-two point groups is given in Table I in both the Schoenflies and Hermann–Mauguin notations. Examples are given of molecules belonging to the more common point groups. In the Hermann–Mauguin description of a point group the minimum number of symmetry elements which will unambiguously specify the complete symmetry is given. We write first the order of the principal rotation or inversion axis n or $\bar{n}$. If there is a twofold axis at right angles to the principal axis, there must be n such twofold axes, and we write $n2$ or $\bar{n}2$. If there is a mirror plane lying in the principal axis, there must be n such planes, and we write nm or $\bar{n}m$. If there is a mirror plane at right angles to the principal axis we write n/m or $\bar{n}/m$. Finally if there are mirror planes both at right angles to, and lying in, the principal rotation axis we write $n/m.m$, which is usually contracted to n/mm.

In the Schoenflies notation rotation axes are regarded as being vertical. A rotation axis of order n with a horizontal plane of symmetry normal to the axis is denoted C_{nh}. As we have seen, the corresponding symbol in the Hermann–Mauguin system is n/m. Thus $C_{2h} \equiv 2/m$. The symbol C_{nv} denotes a vertical rotation axis of order n with n vertical planes of symmetry lying in the rotation axis. The corresponding Hermann-Mauguin symbol is nm. Thus $C_{3v} \equiv 3m$. A rotation axis of order n with n twofold axes normal to it is written D_n. A horizontal mirror plane added to the groups D_n gives D_{nh}. Diagonal vertical planes added to D_n give D_{nd}. Finally, the tetrahedral and octahedral groups are, in the Schoenflies system, labelled T and O respectively.

This may sound rather a lot to swallow all at once. The best thing to do is to select molecules from some of the more common point groups for which examples are given in Table I, and to draw them out, or better still to make three-dimensional models with cork balls and wire, or plasticine and matches. With a little practice of this sort the point groups are quite easy to understand and remember.

TABLE I

The thirty-two crystallographic point groups

Symbol		Plane	Axes of symmetry				Centre	
S	$H-M$	$m(C_s)$	$6(C_6)$	$4(C_4)$	$3(C_3)$	$2(C_2)$	$\bar{1}(C_i)$	Example
C_1	1	..	..	..	..	..	..	CH_3CHO
C_2	2	..	..	..	..	1	..	
C_3	3	..	..	..	1	..	..	
C_4	4	..	..	1	..	..	..	
C_6	6	..	1	..	..	..	..	
C_h	m	1	..	..	..	..	..	HN_3
C_{2h}	$2/m$	1	..	..	..	1	1	$trans$-$CHCl:CHCl$
C_{3h}	$3/m$	1	..	..	1	..	..	
C_{4h}	$4/m$	1	..	1	..	..	1	
C_{6h}	$6/m$	1	1	..	..	..	1	
D_2	222	..	..	..	..	3	..	
D_3	32	..	..	..	1	3	..	
D_4	42	..	..	1	..	4	..	
D_6	62	..	1	..	..	6	..	
D_{2h}	mmm	3	..	..	..	3	1	C_2H_4
D_{3h}	$\bar{6}2m$	4	..	..	1	3	..	BCl_3
D_{4h}	$4/mmm$	5	..	1	..	4	1	
D_{6h}	$6/mmm$	7	1	..	..	6	1	C_6H_6
S_2	$\bar{1}$	..	..	..	..	..	1	
S_4	$\bar{4}$	..	..	..	..	1	..	
S_6	$\bar{3}$	..	..	..	1	..	1	
D_{2d}	$\bar{4}2m$	2	..	..	..	3	..	$CH_2:C:CH_2$
D_{3d}	$\bar{3}m$	3	..	..	1	3	1	
C_{2v}	mm	2	..	..	..	1	..	H_2O
C_{3v}	$3m$	3	..	..	1	..	..	NH_3
C_{4v}	$4mm$	4	..	1	..	..	..	IF_5
C_{6v}	$6mm$	6	1	..	..	..	..	
T	23	..	..	..	4	3	..	
O	43	..	..	3	4	6	..	
T_h	$m3$	3	..	..	4	3	1	
O_h	$m3m$	9	..	3	4	6	1	SF_6
T_d	$\bar{4}3m$	6	..	..	4	3	..	CH_4

Although we have now dealt with the thirty-two crystallographic point groups, we have still to allow for the possibility, in crystals, of symmetry elements involving translation. We shall defer our consideration of these until we deal with the solid state in Chapter VII. For the moment it is enough to remember that many molecules possess symmetry, and a complete determina-

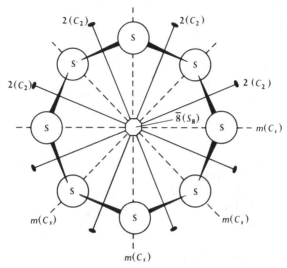

FIG. 22. The cyclic molecule of sulphur, S_8.

tion of the symmetry elements is equivalent to finding the shape of the molecule. Most molecules belong to the thirty-two crystallographic point groups, but there are some important exceptions. Linear molecules belong to the point group ∞/mm ($D_{\infty h}$) if they have a mirror plane at right angles to the internuclear axis as in H_2 (Fig. 1 (a)) and CO_2, or to the point group ∞m ($C_{\infty v}$) if they have no such mirror plane as in HCl (Fig. 1 (b)) and N_2O. Occasionally molecules are found with axes of order five or greater than six. Thus the cyclic S_8 molecule (Fig. 22) found in rhombic and monoclinic sulphur has an eightfold inversion axis and belongs to the point group $\bar{8}2m$ (D_{4d}). Finally, a giant molecule such as a linear polymer may, in the solid state, possess symmetry elements involving translation.

REFERENCES

H. WEYL, *Symmetry* (Princeton University Press).

H. H. JAFFÉ and M. ORCHIN, *Symmetry in Chemistry* (Wiley).

F. A. COTTON, *Chemical Applications of Group Theory* (Wiley).

PART I

Spectroscopic Methods

II

GENERAL CONSIDERATIONS

2.1. Introduction

A MOLECULE consists of an assemblage of atoms arranged in a particular way at a more-or-less fixed distance apart. In this book we shall be concerned with the determination of the relative disposition of these atoms. In other words we are interested in how we can determine the shape and size of a molecule. If we have a triatomic molecule (Fig. 1), we can specify its dimensions

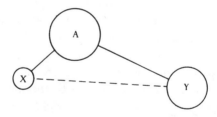

FIG. 1. A triatomic molecule AXY.

by quoting the distances AX, AY, and XY, or we can choose the two distances AX and AY and the angle XAY. In practice we prefer to do the latter. We quote the distances between pairs of atoms joined by a chemical bond and the angles between the bonds. We shall refer to these quantities as the *molecular parameters*. If a polyatomic molecule is made up of N atoms, we have to specify $3N-6$ molecular parameters in order to define the molecule completely, unless the molecule has symmetry in which case the number of parameters will be less. We have here our first illustration of the value of symmetry; it reduces the number of parameters needed to specify the configuration of a molecule. If the symmetry of the water molecule (Fig. 1.18) is mm (C_{2v}) we need only two parameters, because the two O–H distances must be equal. In an octahedral molecule of $m3m$ (O_h) symmetry such as SF_6 (Fig. 2) we need to specify only one para-

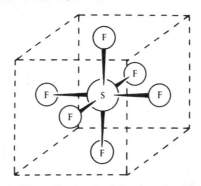

FIG. 2. The molecule of sulphur hexafluoride, SF_6.

meter, the S–F distance, since by symmetry the bond lengths are all equal and all the FSF angles are 90° or 180°. Although the presence of symmetry in a molecule may require the equality of two or more bond lengths, it can never give the actual length of a bond. On the other hand, the presence of symmetry may not only require the equality of two or more angles but may also give their actual magnitudes. This is one reason for expressing molecular parameters in terms of lengths and angles, rather than merely in terms of lengths. A length is always subject to experimental errors but, if symmetry requires an angle to be 90°, then it is 90°.

Some experimental methods yield information only about the shape of a molecule, that is, about the molecular symmetry. On the other hand, some methods may be used to determine the molecular parameters. A determination of all the molecular parameters does, of course, decide the shape of a molecule, but a sharp distinction must be drawn between those methods which yield molecular parameters and those which disclose only molecular symmetry. In Table II we have listed the various methods that we shall discuss, and have shown what type of information each method can give. It will be seen that all these methods can give information about molecular symmetry, but not all can yield molecular parameters.

TABLE II

The capabilities of the various experimental methods used for the determination of molecular structure

Method	Molecular parameters	Molecular symmetry
Pure rotation spectra .	Yes	Yes
Vibration spectra . .	No	Yes
Rotational Raman spectra .	Yes	Yes
Vibrational Raman spectra .	No	Yes
Electron diffraction . .	Yes	Yes
X-ray diffraction . .	Yes	Yes
Neutron diffraction . .	Yes	Yes
Classical stereochemistry .	No	Yes
Dipole moments . .	No	Yes
Magnetic measurements .	No	Yes
Nuclear magnetic resonance .	Yes	Yes

Before we can begin our discussion of the use of spectroscopy for the determination of molecular structure, there are certain aspects of molecules and of spectra that we must consider. Some of these topics are probably familiar, but they are included for the sake of completeness. Others may seem a little abstruse and irrelevant at the moment, but we will deal with them here in order to avoid breaking the thread of the argument in later chapters. It will probably be desirable to refer back to certain portions of this chapter later on.

2.2. Energy levels

It is a basic postulate of the quantum theory that the energy levels of atoms and molecules are discrete. Spectra arise from the emission or absorption of definite quanta of radiation when transitions occur between certain energy levels. Fig. 3 represents diagrammatically some energy levels of a molecule, although for the moment we shall not specify what these energy levels are. The vertical axis is divided into equal arbitrary units of energy.

A transition from the state marked A to that marked B will involve the absorption of a quantum of radiation. If ΔE is the energy difference between the states A and B, the frequency of the quantum is given by Planck's relation

$$\nu = \Delta E/h,$$

where h is Planck's constant and is equal to $6 \cdot 62517 \times 10^{-27}$ erg s. A transition from state D to state C will involve the

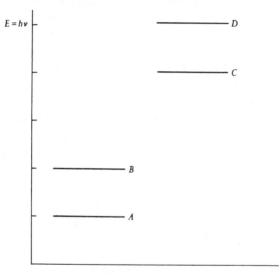

FIG. 3. Some energy levels of a molecule.

emission of a quantum of energy. Even though the absolute energy of the molecule is different in all four states, the magnitudes of the two quanta will be the same and the frequencies associated with the two quanta will be identical, because the energy difference between the states A and B is the same as that between C and D. Spectroscopic methods are concerned with the detection of these quanta, and consequently a spectroscopic method detects the difference in energy between two energy levels.

In an atom the energy levels represent different allowed states for the extranuclear electrons. A molecule also can absorb or emit energy as a result of transitions between different electronic energy levels. However, there are two other principal ways in which a molecule can change its energy, and which cannot occur with atoms. A molecule can absorb a quantum of energy and increase its vibrational energy, or it can absorb a quantum of energy and increase its rotational energy. The quanta associated with these three different types of energy levels—electronic, vibrational, and rotational—are of very different energy. We can, as a good approximation, treat them independently, and consider the total energy E_T to be made up of three parts

$$E_T = E_{el} + E_{vib} + E_{rot}.$$

In general the order of magnitude of the difference between electronic levels is $\sim$100 kcal/mole; that between vibrational levels is $\sim$5 kcal/mole; and that between rotational levels is $\sim$ 0·01 kcal/mole. Since the average thermal kinetic energy of a molecule at room temperature is $\sim$ 1 kcal/mole, it can be seen from the Boltzmann distribution law that molecules in general will be in the lowest vibrational level of the ground electronic state, but will possess several quanta of rotational energy. Transitions between different electronic levels give rise to spectra in the visible or ultra-violet region of the electromagnetic spectrum, and these spectra are referred to as *electronic spectra*. Transitions between vibrational levels within the same electronic level give rise to spectra which are called vibration spectra, in the near infra-red. Transitions between different rotational levels within the same vibrational level are responsible for spectra in the far infra-red or microwave region, and these spectra are called rotation spectra. In fact a transition between two electronic levels is usually accompanied by changes in the vibrational and rotational quantum numbers, so that electronic spectra are really electronic-vibration-rotation spectra. In the same way, vibration spectra are really *vibration-rotation spectra* and are normally given this name. The spectra corresponding to the smallest changes in energy are then called *pure rotation spectra*.

A diagram illustrating the distribution of the different energy levels is shown in Fig. 4. Two electronic levels E'' and E' are shown. The lower level E'' has a number of vibrational levels denoted by the vibrational quantum number $v'' = 0, 1, 2,...,$ and each vibrational level has its own rotational levels represented by the rotational quantum number $J'' = 0, 1, 2,... .$

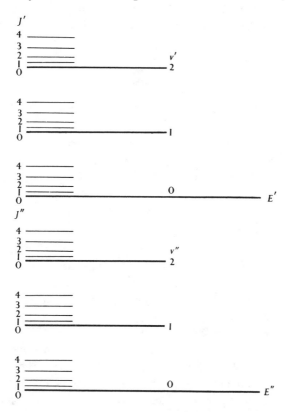

Fig. 4. Electronic, vibrational, and rotational energy levels of a diatomic molecule.

In the same way the upper electronic level E' has vibrational levels $v' = 0, 1, 2,...,$ and each vibrational level has rotational levels $J' = 0, 1, 2,... .$

2.3. Potential energy curves

Some of the ideas discussed in the previous section are perhaps made clearer by a consideration of the potential energy curve of a diatomic molecule. Such a molecule consists of two positively charged nuclei and a number of electrons. The stable molecule corresponds to a minimum in the potential energy of the system. The minimum is a result of the balance between the repulsion of the two positively charged nuclei and the attraction of the electronic binding. The repulsion comes into play at small

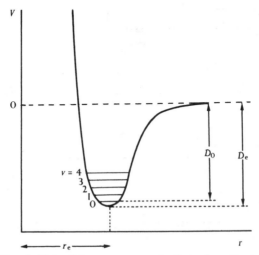

FIG. 5. Potential energy curve of a diatomic molecule.

values of the internuclear distance r, but increases very rapidly as r is diminished still further. The attraction depends less strongly on r and operates over greater distances. The resultant of these two forces is shown diagrammatically in Fig. 5. The attractive force approaches zero asymptotically as r increases towards infinity. We define our zero of potential energy as the potential energy of the two atoms separated by an infinite distance. For practical purposes even the attractive force operates over a distance of only a few Ångström units ($1 \text{ Å} = 10^{-8}$ cm).

We consider one atom to be fixed on the vertical axis while the other atom vibrates about the minimum of the curve, the internuclear distance changing in accordance with the curve. The vibrational energy is quantized, and this is represented by the horizontal lines which are labelled with the vibrational quantum number $v = 0, 1, 2,\ldots$. For small movements of the atoms from their equilibrium position, defined by the minimum of the curve, the restoring force is proportional to the displacement and the atoms vibrate with simple harmonic motion in accordance with Hooke's law. The lower portion of the curve is, therefore, parabolic in shape. At higher vibrational energies it becomes more difficult to compress the molecule than to extend it, the curve departs from a parabola, and the vibration is said to become *anharmonic*. At sufficiently high vibrational energies the molecule will dissociate.

It should be noticed that the energy at the minimum of the curve is not the same as that in the ground vibrational state $v = 0$. In other words, even though all the quanta of vibrational energy have been removed from the molecule by cooling it to the absolute zero of temperature, the molecule is still vibrating. This is a consequence of Heisenberg's uncertainty principle which states that

$$\Delta x \Delta p \sim h,$$

where Δx is the uncertainty in the position of a particle and Δp is the uncertainty in the momentum. Thus the only way in which the position can be completely specified is if the momentum is infinite, and the only way in which the momentum can be completely defined is if the position is infinitely variable. At the minimum of the curve the internuclear distance is fixed and the momentum is zero, so that the uncertainty principle is violated on both counts. We must conclude, therefore, that the minimum represents a hypothetical state of the molecule, and that in the state of lowest vibrational energy the molecule is vibrating. The energy difference between the minimum of the curve and the lowest vibrational state for $v = 0$ is called the *zero-point energy of vibration*. The hypothetical internuclear distance corresponding to the minimum of the curve is called the *equilibrium internuclear distance* r_e. The height of the asymptote from the minimum gives the *spectroscopic heat of dissociation* D_e. The *chemical heat of dissociation* D_0 is the height of the asymptote from the ground vibrational state at $v = 0$.

2.4. Equilibrium internuclear distance r_e and apparent internuclear distance r_0

Although it is possible, as we shall see in section 3.6, in certain simple cases to obtain a value for the equilibrium internuclear distance r_e, the distance normally determined is the internuclear distance in the ground vibrational state, since most molecules are in this state at room temperature. This leads to a difficulty because the molecule is vibrating in the lowest vibrational state, and hence the internuclear distance is constantly changing. Let us call the instantaneous distance between two nuclei of a molecule in the ground vibrational state r, and suppose, for the moment, that we label as r_0 a quantity that we can call 'the internuclear distance in the ground vibrational state'. Clearly, in order to give a precise expression for r_0 we must define it in some way as an average value of r. The question now is: What particular average shall we choose? At first sight it might seem that everybody ought to get together and agree on a particular average. But the problem is not simply one of nomenclature, and there is a more fundamental difficulty than that. The theory behind each experimental method necessarily leads to the selection of an average, and not all these averages are the same. For instance, spectroscopic methods determine an average value of $1/r^2$, so that the spectroscopic average must be defined as

$$\frac{1}{r_0} = \left(\frac{1}{r^2}\right)^{\frac{1}{2}}_{\mathrm{av}}.$$

The theory of nuclear magnetic resonance experiments on solids tells us that what we determine is really an average value of $1/r^6$, so that the magnetic resonance average must be

$$\frac{1}{r_0} = \left(\frac{1}{r^6}\right)^{\frac{1}{6}}_{\mathrm{av}}.$$

To make matters worse, the distance determined by X-ray diffraction is not an internuclear distance at all, but the time-averaged distance between the centres of gravity of two electronic clouds. Nevertheless, the final result of all these methods is quoted as an r_0 value in order to distinguish r_0 from r_e. Fortunately there are good reasons for believing that, whatever experimental method is employed, the values obtained for r_0 will not differ by much. However, we must remember that each experimental technique defines r_0 in a slightly different way, and when we compare the same bond length determined by two different techniques we are really comparing two slightly different quantities. In order to emphasize that there is this element of arbitrariness in r_0 we shall refer to it as the *apparent internuclear distance*.

2.5. Infra-red and Raman spectroscopy

Molecular spectra are usually observed in absorption. For instance, in the infra-red a collimated beam of radiation from a body at red heat is passed through a sample of the substance under investigation and then dispersed by a prism or grating. The prism must be transparent to the radiation, and in the infra-red sodium chloride or potassium bromide prisms and windows are usually used. The dispersed beam then passes to a detector which looks at each small portion of the spectrum separately. The smaller the portion of the spectrum that falls on the detector at any one moment, the greater is the *resolution* of the instrument. The detector finds that certain parts of the continuous spectrum of the source have been weakened relative to others, and the frequencies of these attenuated parts correspond to the absorption frequencies of the molecule.

We have, however, to consider a second sort of effect, the Raman effect, which is frequently complementary to that described above. In Raman spectroscopy a monochromatic beam of light of any convenient wavelength is shone on to the sample, and observations are made on the scattered light at right angles to the incident beam. If the scattered light is dispersed by means of a prism or, for higher resolution, a grating, a spectrum consisting of discrete lines may be observed. Suppose a quantum of frequency ν_0 and energy $h\nu_0$ collides with a molecule of the sample gas. The light quantum may be scattered with unchanged frequency and it will then form part of the *Rayleigh line*. On the other hand the incident quantum may induce a transition in the sample molecule. Suppose, for example, that this is a vibrational transition from the state $v = 0$ to the state $v = 1$. The quantum associated with this transition will have a frequency which we shall call ν_{v}, and an energy $h\nu_{\mathrm{v}}$. Since the incident quantum has induced this transition it will be scattered with a diminished energy $h(\nu_0 - \nu_{\mathrm{v}})$. In other words a *Raman line* will be observed on the low-frequency side of the Rayleigh line with a frequency shift of ν_{v}. Such a line is known as a *Stokes line*. If the molecule happens to be initially in a state $v = 1$, the transition $v = 1$ to $v = 0$ may be induced as a result of the collision with the incident quantum. In this case the quantum will be scattered with an enhanced energy $h(\nu_0 + \nu_{\mathrm{v}})$ and a line will appear on the high-frequency side of the Rayleigh line with the same frequency shift as before. Such a line is called an *anti-Stokes line*. In general the transition may be either rotational or vibrational and in each case the Raman lines will be observed at frequencies $\nu_0 \pm \nu_{\mathrm{R}}$, where ν_{R} is a particular Raman frequency and corresponds to a particular rotational or vibrational transition.

Since, as we have seen, the majority of molecules will be in the lowest vibrational state at room temperature, the Stokes lines will be stronger than the anti-Stokes lines in the vibration spectrum. In the pure rotation spectrum the intensity relations are more complicated because the molecules may possess several quanta of rotational energy at the temperature of the experiment.

2.6. The uses of spectroscopy

Table III shows the three types of spectra that are observed and the quantities that may be found from investigations in each region of the electromagnetic spectrum. Table III emphasizes some interesting points about spectra. Firstly, if the rotational fine structure can be analysed, molecular parameters may be obtained, and this may be achieved in all three relevant regions. Secondly, information about molecular symmetry may also be obtained from all three portions of the spectrum. Finally an investigation of the electronic spectrum does not give us any information about molecular symmetry or molecular parameters that cannot already be obtained from the pure rotation or the

TABLE III

The uses of emission or absorption spectroscopy for the determination of molecular structure

Spectrum	Energy level differences that may be observed	Quantities that may be determined
{ Pure rotation { Microwave	Rotational	Molecular symmetry Molecular parameters
Vibration- rotation	Rotational	Molecular symmetry Molecular parameters
	Vibrational	Molecular symmetry
Electronic	Rotational	Molecular symmetry Molecular parameters
	Vibrational	Molecular symmetry
	Electronic	..

vibration-rotation spectra. Actually this last statement is not quite true because it is possible, from the electronic spectrum, to obtain the symmetry or the parameters of a molecule in an excited electronic state: but we shall not be concerned with electronically excited molecules. In this book, therefore, we shall not discuss electronic spectra, because we shall already have dealt with the relevant principles in the next two chapters on pure rotation and vibration-rotation spectra. It would be wrong to conclude from this omission that ultra-violet spectroscopy is unimportant. It can give information about many things, such as the composition of large molecules, the relation between colour and constitution, the size and shape of molecular fragments, the electronic structure of molecules, and the shape of potential energy curves, things that can seldom be determined from pure rotation or vibration-rotation spectra. Nevertheless, because this book is restricted to the determination of the size and shape of ordinary molecules, we shall have little more to say about electronic spectra.

The uses of Raman spectroscopy are similar to those of absorption spectroscopy. If the rotational Raman lines can be resolved, molecular parameters may be determined. If, however, only vibrational Raman lines are observed, information may be obtained only about molecular symmetry. There is nothing in the Raman effect that corresponds to electronic spectra in absorption or emission spectroscopy.

2.7. Selection rules

From what has been said, it might appear that transitions between all possible energy levels may occur. In fact this is not so. In just the same way that an atom may change from an S to a P, or a P to a D, but not from an S to a D state, so there are restrictions on the allowed changes in molecules. These restrictions are known as *selection rules*. We shall distinguish two sorts of selection rules. The conditions that govern changes in the individual quantum numbers, as in the example given above for atoms, we shall call *particular selection rules*, and we shall discuss them when we consider the various types of spectra. There are also *gross selection rules* which can be used to decide whether a molecule will give a vibration or rotation spectrum at all. Thus all molecules have an electronic spectrum, so that there are no gross selection rules governing spectra in the visible or ultra-violet. On the other hand, a homonuclear diatomic molecule, such as Cl_2, cannot exhibit a vibration spectrum in the infra-red, and we have here an example of a gross selection rule. Both types of selection rule are based on symmetry, and there is no fundamental difference between them. Nevertheless, we shall find it convenient to make the distinction.

If we restrict ourselves to diatomic molecules it is not difficult to obtain a qualitative understanding of gross selection rules. When a heteronuclear molecule, such as HCl, vibrates, there is a net oscillatory displacement of charge. The molecule behaves like an isolated vibrating charge, and can interact with the oscillatory electric component of any incident radiation. The vibration is *allowed* and will be active in the infra-red. However, in a homonuclear diatomic molecule, such as Cl_2, symmetry demands that every displacement in one half of the molecule shall be balanced by an equal but opposite change in the other half. Thus there is no net displacement of charge and hence no interaction with the incident radiation. The vibration is *forbidden* and will be inactive in the infra-red. Actually, because of the finite velocity of light and hence of the finite time required for a disturbance in one half of the molecule to reach the other half, no transition is absolutely forbidden, and all vibrational and rotational motions interact with the incident radiation to some extent. However, the intensity of allowed transitions is about 10^8 times as great as forbidden ones, and we are to this extent justified in the use of the word forbidden. Anything that destroys the symmetry of a molecule, such as a collision, will bring about a violation of the selection rules based on symmetry, although the intensity of the transitions will still remain small.

For polyatomic molecules a similar argument applies, except that there will be several vibrations to be considered instead of only one. Each vibrational frequency can be associated with a different motion of the nuclei. For a vibration to be active in the infra-red, the motion of the nuclei must produce a change in the dipole moment of the molecule.† Except for diatomic molecules, this does not, of course, require the molecule to have a permanent dipole moment since the vibration may change the dipole moment from a zero value to a finite one. We may express this gross selection rule by saying that, for a frequency to appear in the infra-red,

$$\frac{d\mu}{dq} \neq 0,$$

where μ is the dipole moment of the molecule and q represents either one coordinate or several coordinates defining the motion of the nuclei.

In a similar way, for a molecule to show a rotation spectrum in the far infra-red or microwave region, the dipole moment of the molecule must change with respect to the direction of the incident radiation as the molecule rotates. The molecule must, therefore, possess a permanent dipole moment. The gross selection rule for the existence of a pure rotation spectrum is therefore

$$\mu \neq 0.$$

In the Raman effect the gross selection rule for the vibration

† Dipole moments are discussed in section 11.1.

spectrum of a molecule states that the motion of the nuclei must produce a change in polarizability of the molecule,† i.e.

$$\frac{d\alpha}{dq} \neq 0,$$

where α is the polarizability of the molecule. Since the polariz-ability of a homonuclear diatomic molecule does change as the length of the bond is altered, such a molecule will give a vibra-tional Raman line. Finally, for a molecule to show a rotational Raman spectrum, the polarizability of the molecule perpendicu-lar to the axis of rotation must be different in different direc-tions so that the polarizability will change with respect to the direction of the incident radiation as the molecule rotates. Any molecule with a rotation axis of order three or more will show no rotational Raman lines as a result of rotation about the principal axis, since the polarizability will be identical in every direction at right angles to this axis. Thus benzene (Fig. 1.20) will show no rotational Raman spectrum as a result of rotation round the sixfold axis, but only as a result of rotation about axes in the plane of the molecule.

These gross selection rules are summarized in Table IV.

TABLE IV

The gross selection rules governing activity in the Raman and infra-red

Spectrum		Gross selection rule
Vibrational infra-red	.	Vibration must involve change in dipole moment of molecule
Vibrational Raman .	.	Vibration must involve change in polarizability of molecule
{ Rotational infra-red Microwave	.	Molecule must possess permanent dipole moment
Rotational Raman .	.	Polarizability perpendicular to axis of rotation must be anisotropic.

2.8. Units and nomenclature

The electromagnetic spectrum covers an enormous range of frequencies, and spectroscopists working in the different regions have found it convenient to use their own units. These units

---

† Polarizabilities are discussed in section 11.2.

have usually been selected in such a way as to make the numbers of reasonable size and to avoid the inclusion of large powers of ten. In the X-ray, ultra-violet, and visible regions the spectroscopists refer to the wavelength of the radiation and measure the wavelength in Ångström units (1 Å = 10^{-8} cm). In the near and far infra-red microns are employed as the unit of wavelength

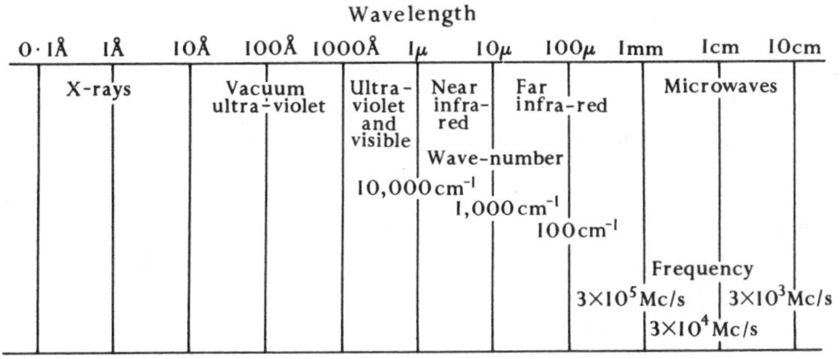

FIG. 6. The electromagnetic spectrum.

(1 μ = 10^{-4} cm). However, in the infra-red it is often convenient to work in terms of wave-numbers, that is the number of wavelengths per cm. The wave-number is given by the true frequency divided by the velocity of light, i.e.

$$\text{Wave-number} = \frac{\text{frequency}}{\text{velocity of light}} = \frac{s^{-1}}{cm\ s^{-1}} = cm^{-1}.$$

The units of wave-numbers are therefore reciprocal centimetres or 'centimetres-to-the-minus-one'. Unfortunately, wave-numbers are often referred to as frequencies, and statements such as 'X–H bonds have a frequency of about 3,000 wave-numbers' are very common. Moreover, the same symbol ν is often given to both wave-numbers and frequencies, although occasionally the symbol $\tilde{\nu}$ is used for wave-numbers. The context usually removes any ambiguity, but it is important to remember to convert to true frequencies in any calculations. Microwave spectroscopists deal with true frequencies and express their results in megacycles per second (1 Mc/s = 10^6 c/s) or in kilocycles per second (1 kc/s = 10^3 c/s).

The electromagnetic spectrum from X-rays to microwaves is shown in Fig. 6. The wavelength scale is logarithmic, and other commonly employed units are included in the appropriate regions.

REFERENCE

F. O. RICE and E. TELLER, *The Structure of Matter* (Wiley).

III

PURE ROTATION SPECTRA

3.1. Diatomic molecules

THE investigation of pure rotation spectra provides a powerful method of finding molecular parameters of simple molecules that may be obtained in the gas phase. In order to show a pure rotation spectrum at all, the molecule must obey the gross selection rule: that is, the molecule must possess a permanent dipole moment.

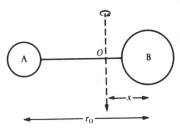

FIG. 1. A diatomic molecule AB.

As a suitable model for the interpretation of pure rotation spectra we may consider two atoms A and B of mass m_A and m_B at a fixed distance r_0 apart (Fig. 1). This is known as the *rigid-rotor model*. If this molecule rotates about an axis perpendicular to the internuclear axis through the centre of gravity O, the moment of inertia is given by

$$I_0 = m_A(r_0-x)^2 + m_B x^2,$$

where x is the distance from atom B to the centre of gravity. If we take moments about the centre of gravity, we get

$$m_A(r_0-x) = m_B x.$$

Therefore
$$x = \frac{m_A r_0}{m_A + m_B}.$$

Hence
$$I_0 = m_A\left(r_0 - \frac{m_A r_0}{m_A+m_B}\right)^2 + m_B\left(\frac{m_A r_0}{m_A+m_B}\right)^2 = \left(\frac{m_A m_B}{m_A+m_B}\right)r_0^2. \quad (1)$$

The reduced mass M of the molecule is defined as

$$\frac{1}{M} = \frac{1}{m_A} + \frac{1}{m_B}.$$

Hence

$$M = \frac{m_A m_B}{m_A + m_B}.$$

Thus, from equation (1),

$$I_0 = M r_0^2.$$

This means that we can regard our molecule spinning about its centre of gravity as equivalent to a single particle of mass M describing a circle of radius r_0. The motion of a particle in a circle is a well-known problem in wave mechanics and an exact solution of the Schrödinger wave equation can be obtained. It is found, as usual, that only certain discrete values of the energy of the system are possible, and these values are given by

$$E_{\text{rot}} = \frac{h^2 J(J+1)}{8\pi^2 M r_0^2} = \frac{h^2 J(J+1)}{8\pi^2 I_0} = B_0 h J(J+1),$$

where

$$B_0 = h/8\pi^2 I_0. \tag{2}$$

B is called the *rotational constant*, the subscript denoting to which vibrational state the constant refers. The rotational quantum number J can have integral values including zero. If $J = 0, E_{\text{rot}} = 0$ and there is no rotational energy. In other words there is no zero-point energy of rotational motion. Rotation differs from vibration in this respect because, if all the momentum is removed from a rotor, the position is still infinitely variable. Heisenberg's principle is thus not disobeyed.

There is an element of inconsistency in the above derivation in that we have stated that the atoms are at a fixed distance apart but we have called this distance r_0. We have adopted this notation because the above formulae can be applied to real molecules, and the distance that is then obtained is the apparent internuclear distance r_0, and not the equilibrium internuclear distance r_e.

It can be shown that the particular selection rule applicable to the pure rotation spectrum of a diatomic molecule is

$\Delta J = \pm 1$. The energy difference between two successive levels is given by

$$\Delta E_{\text{rot}} = B_0 h\{J'(J'+1) - J''(J''+1)\}, \tag{3}$$

where J' refers to the upper rotational state and J'' to the lower. Since the particular selection rule demands that $J' - J'' = 1$,

$$\Delta E_{\text{rot}} = B_0 h 2J'.$$

From this equation we obtain the energy level differences shown in Table V. Although the spacing between the energy levels increases linearly with J, the spectrum will consist of a number of equally spaced lines as shown in Fig. 2. Since $\nu = E/h$, the frequency separation will be

$$\Delta \nu = 2B_0.$$

TABLE V

The pure rotational levels of a rigid rotor in the infra-red spectrum

$E_{\text{rot}} = B_0\,hJ(J+1)$	$\Delta E_{\text{rot}} = B_0\,h2J'$	J
0		0
$2B_0\,h$	$2B_0\,h$	1
$6B_0\,h$	$4B_0\,h$	2
$12B_0\,h$	$6B_0\,h$	3
$20B_0\,h$	$8B_0\,h$	4
$30B_0\,h$	$10B_0\,h$	5
$42B_0\,h$	$12B_0\,h$	6

Thus from the spacing we may obtain the moment of inertia directly from equation (2) and, knowing the masses, we may calculate the sole molecular parameter r_0 from equation (1).

As an illustration of the application of the rigid-rotor approximation we may consider the far infra-red spectrum of HI.† The rotation spectrum is found to consist of a series of lines with a separation $\Delta\tilde{\nu} = 12\cdot 8$ cm^{-1}. The true frequency of rotation is therefore

$$\Delta\nu = c/\lambda = (3\cdot 00 \times 10^{10})(12\cdot 8) = 3\cdot 84 \times 10^{11} \text{ c/s}.$$

The energy difference between two successive levels is

$$h\Delta\nu = (6\cdot 62 \times 10^{-27})(3\cdot 84 \times 10^{11}) = 2\cdot 54 \times 10^{-15} \text{ ergs}.$$

† M. Čzerny, *Z. Phys.* 1927, **44**, 235.

Now

$$\Delta E_{\text{rot}} = h\Delta\nu = h2B_0 = \frac{2h^2}{8\pi^2 I_0} = 2\cdot54 \times 10^{-15} \text{ ergs.}$$

Hence $$I_0 = 4\cdot37 \times 10^{-40} \text{ g cm}^2.$$

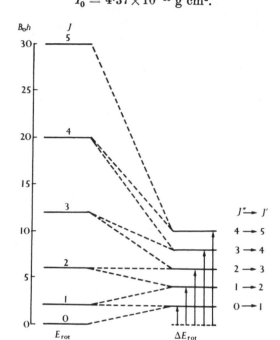

FIG. 2. Pure rotation spectrum of a heteronuclear diatomic molecule.

Iodine consists of a single isotope of mass number 127, so that the reduced mass of the HI molecule is given by

$$M = \frac{127 \times 1}{127+1} \times \frac{1}{6\cdot02 \times 10^{23}} = 1\cdot65 \times 10^{-24} \text{ g.}$$

Therefore

$$r_0^2 = \frac{4\cdot37 \times 10^{-40}}{1\cdot65 \times 10^{-24}} = 2\cdot65 \times 10^{-16} \text{ cm}^2,$$

and $$r_0 = 1\cdot63 \text{ Å.}$$

A close examination of the lines in the pure rotation spectrum shows that in fact the spacing between the lines decreases at higher values of J. The reason for this is that, in the higher rotational levels, the molecule is spinning faster and the bond is stretched slightly by a centrifugal force. In this way the moment of inertia is increased and consequently the spacing is decreased. In order to fit the energy levels it is found that a correction term is needed, and the energy of the levels must be written

$$E_{\text{rot}} = B_0\,hJ(J+1) - D_0\,h\{J(J+1)\}^2.$$

The constant D is called the *centrifugal distortion constant* and is of the order of one ten-thousandth part of B. For low values of J the second term is negligible. The centrifugal distortion constant in the ground vibrational state and the chemical heat of dissociation are both given the same symbol D_0, but there is little opportunity for confusion.

3.2. Linear polyatomic molecules

The formulae for the energy levels given in section 3.1 apply also to linear polyatomic molecules. The far infra-red spectrum is a difficult portion of the electromagnetic spectrum in which to work, and pure rotation spectra are better investigated in the microwave region. Microwave spectroscopy is a very useful technique in that it can provide values of molecular dipole moments, and nuclear spins, moments, and quadrupole moments. However, we are not concerned with these quantities just now. For our purposes there is little difference between microwave spectroscopy and far infra-red spectroscopy, except that it is much easier to obtain a practically monochromatic beam of microwaves. Microwave spectroscopy is therefore capable of much higher resolution and enables very accurate values to be obtained for rotational constants and molecular parameters. We may illustrate a typical application by a consideration of the linear triatomic molecule carbonyl sulphide, OCS.†

We are immediately up against a difficulty. OCS has two bond lengths but only one moment of inertia. We can solve this problem by working with two isotopic molecules and assuming that the bond lengths are the same in the two species. We may thus obtain two moments of inertia and then solve for the two molecular parameters.

† C. H. Townes, A. N. Holden, and F. R. Merritt, *Phys. Rev.* 1948, **74**, 1113

The species $^{16}O^{12}C^{32}S$ and $^{16}O^{12}C^{34}S$ have been investigated in the microwave region and the measured frequencies of the lines are shown in Table VI. These frequencies were obtained with an accuracy of four parts in 10^7. From these results values of the rotational constant B_0 and the centrifugal distortion

TABLE VI

The observed microwave spectrum of $^{16}O^{12}C^{32}S$ and $^{16}O^{12}C^{34}S$

Transition	$^{16}O^{12}C^{32}S$	$^{16}O^{12}C^{34}S$
$1 \to 2$	24325·92 Mc/s	23732·33 Mc/s
$2 \to 3$	36488·82	
$3 \to 4$	48651·64	47462·40
$4 \to 5$	60814·08	

TABLE VII

The rotational constants, centrifugal distortion constants, and moments of inertia of $^{16}O^{12}C^{32}S$ and $^{16}O^{12}C^{34}S$

Molecule	B_0	D_0	I_0
$^{16}O^{12}C^{32}S$	6081·480 Mc/s	1600 kc/s	$137·974 \times 10^{-40}$ g cm^2
$^{16}O^{12}C^{34}S$	5932·843	1400	141·431

constant D_0 were obtained for each of the two species. These values are given in Table VII. If we now take $h = 6·62517 \times 10^{-27}$ erg s, we obtain the moments of inertia shown in the last column of Table VII. Using the notation in Fig. 3, we may write immediately for the moment of inertia of $OC^{32}S$ about an axis perpendicular to the internuclear axis through the centre of gravity,

$$^{32}I_0 = m_{^{16}O} \times a^2 + m_{^{12}C} \times b^2 + m_{^{32}S} \times c^2. \tag{4}$$

In order to write down the corresponding moment of inertia of $OC^{34}S$ it is convenient to use the theorem of parallel axes, which states that the moment of inertia of a rigid body referred to an arbitrary axis is equal to the moment of inertia referred to a parallel axis through the centre of gravity plus the moment of inertia of the entire mass about the given axis, assuming it to be concentrated at the centre of gravity. In our case the arbitrary given axis passes through the point that corresponds to the centre of gravity of $OC^{32}S$, and the centre of gravity of $OC^{34}S$ is moved by a distance d from this point towards the ^{34}S atom (Fig. 3). We obtain

$$^{34}I_0 + {}^{34}Md^2 = m_{^{16}O}\, a^2 + m_{^{12}C}\, b^2 + m_{^{34}S}\, c^2, \qquad (5)$$

where ^{34}M is the total mass of OC^{34}S.

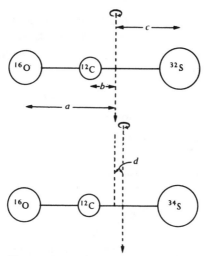

FIG. 3. The molecules OC^{32}S and OC^{34}S.

We may obtain an expression for d by taking moments about the centre of gravity of each molecule,

$$m_{^{16}O}\, a + m_{^{12}C}\, b = m_{^{32}S}\, c, \qquad (6)$$

$$m_{^{16}O}(a+d) + m_{^{12}C}(b+d) = m_{^{34}S}(c-d). \qquad (7)$$

Equation (7) gives

$$d(m_{^{16}O} + m_{^{12}C} + m_{^{34}S}) = {}^{34}Md = m_{^{34}S}\, c - m_{^{16}O}\, a - m_{^{12}C}\, b.$$

Using equation (6) we get

$$^{34}Md = m_{^{34}S}\, c - m_{^{32}S}\, c,$$

therefore

$$d = \left(\frac{m_{^{34}S} - m_{^{32}S}}{{}^{34}M}\right)c. \qquad (8)$$

If we substitute this value of d into equation (5), we obtain for the moment of inertia of OC^{34}S

$$^{34}I_0 = m_{^{16}O}\, a^2 + m_{^{12}C}\, b^2 + \left\{ m_{^{34}S} - \frac{(m_{^{34}S} - m_{^{32}S})^2}{{}^{34}M} \right\}c^2. \qquad (9)$$

If we now subtract (9) from (4), and substitute $m_{16_O} = 16 \cdot 00000$, $m_{12_C} = 12 \cdot 00386$, $m_{32_S} = 31 \cdot 98089$, $m_{34_S} = 33 \cdot 97711$, and $N_0 = 6 \cdot 02486 \times 10^{23}$ mole^{-1}, we obtain a value for $c = 1 \cdot 0383$ Å. We may substitute this value of c into (8) and obtain $d = 0 \cdot 0334$ Å.

Next we substitute our value of c into equation (6) and obtain an expression for a in terms of b. We substitute this value of a into (4), and we then have a quadratic to solve for b. We find $b = 0 \cdot 5214$ Å and thence $a = 1 \cdot 6842$ Å. Thus the bond lengths are

$$r_0(CO) = a - b = 1 \cdot 1628 \text{ Å},$$

$$r_0(CS) = b + c = 1 \cdot 5597 \text{ Å}.$$

From an investigation of several different isotopic species the results shown in Table VIII were obtained. It will be seen that our results differ slightly from those given in this table. This is because we have used more recent values for the fundamental

TABLE VIII

Internuclear distances obtained from various combinations of different isotopic species of the same molecule

Isotopic species		$r_0(CO)$	$r_0(CS)$
$^{16}O^{12}C^{32}S$	$^{16}O^{12}C^{34}S$	$1 \cdot 1647$ Å	$1 \cdot 5576$ Å
$^{16}O^{12}C^{32}S$	$^{16}O^{13}C^{32}S$	$1 \cdot 1629$	$1 \cdot 5591$
$^{16}O^{12}C^{34}S$	$^{16}O^{13}C^{34}S$	$1 \cdot 1625$	$1 \cdot 5594$
$^{16}O^{12}C^{32}S$	$^{18}O^{12}C^{32}S$	$1 \cdot 1552$	$1 \cdot 5653$

constants. We can see that uncertainties in the values of h and N_0 (mainly in h) can result in errors of the order of $0 \cdot 002$ Å in the bond lengths. An error of $0 \cdot 001$ Å may be present because of uncertainties in the atomic masses. The remainder, and the largest part, of the discrepancies in Table VIII is due to the effect of zero-point vibration. Only if we were measuring r_e values would we expect the internuclear distances in the different isotopic species to be the same. We shall meet this point again later on (section 3.6).

Isotopic substitution has helped a great deal in the determination of internuclear distances. The high sensitivity and resolution of microwave spectroscopy enables the transitions of many of the numerous isotopic species to be observed in their natural abundances. In other regions of the electromagnetic spectrum, however, only the substitution of deuterium for hydrogen has proved of much value since other isotopic substitutions generally involve too small a proportional change in mass to allow resolution. It should be noted that only one substitution at a given atomic position gives additional independent equations. Nevertheless, further substitutions are valuable since they give us an estimate of the accuracy of the molecular parameters obtained.

A selection of bond lengths of linear polyatomic molecules, obtained by the method described in this section, is given in Table IX.

TABLE IX

Values obtained by microwave spectroscopy of the bond lengths of some selected linear polyatomic molecules†

Molecule	Bond	r_0	Bond	r_e
HCN	CH	1·064 Å	CN	1·156 Å
NNO	NN	1·126	NO	1·191
ClCN	CCl	1·629	CN	1·163
BrCN	CBr	1·790	CN	1·159
ClCCH	CH	1·052	CC	1·211
	CCl	1·632		
HC≡C—CN	CH	1·057	CN	1·157
	C—C	1·382	C≡C	1·203

3.3. Symmetric tops

A *symmetric top molecule* is one which has three finite moments of inertia of which two are equal. Any molecule which possesses a single axis of rotation of order three or more is a symmetric top. A typical example is CH_3Cl (Fig. 4), and the majority of symmetric tops so far investigated have this $3m$ (C_{3v}) symmetry. We shall suppose that the a axis coincides with the unique axis. We then have $I^a \neq I^b = I^c$. The wave equation for a rigid symmetric top has been set up and solved to give, for the energy levels,

$$E_{\text{rot}} = \frac{h^2}{8\pi^2 I^b} J(J+1) + \frac{h^2}{8\pi^2}\left(\frac{1}{I^a} - \frac{1}{I^b}\right)K^2, \qquad (10)$$

or $\qquad E_{\text{rot}} = B_0^b h J(J+1) + h(B_0^a - B_0^b)K^2,$

where $\qquad B_0^a = h/8\pi^2 I_0^a$ and $B_0^b = h/8\pi^2 I_0^b.$

† W. Gordy, W. V. Smith, and R. F. Trambarulo, *Microwave Spectroscopy.*

We have had to introduce two rotational quantum numbers J and K, since there are two types of rotation in a symmetric top. Both these quantum numbers can have any integral value, except that K cannot be greater than J. The quantum number

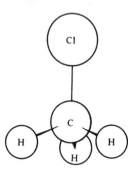

FIG. 4. The molecule of methyl chloride, CH_3Cl.

K refers to rotation round the unique axis, and J refers to an end-over-end rotation of the molecule. A rotation of a symmetric top about the unique axis will cause no change in dipole moment with respect to the direction of the incident radiation, so that no absorption will result from this type of rotation. We are not surprised to find, therefore, that the particular selection rule for K is $\Delta K = 0$. The particular selection rule for J is, as for linear molecules, $\Delta J = \pm 1$. Since K cannot change, the last term in equation (10) will disappear when a difference is taken, and we find for the energy difference between successive rotational levels of a symmetric top

$$\Delta E_{\text{rot}} = B_0^b h\{J'(J'+1) - J''(J''+1)\}$$

with $J' - J'' = 1$. This formula is identical with equation (3) which refers to linear molecules, and the treatment of symmetric tops follows exactly our treatment of linear molecules. Thus only one moment of inertia is obtainable from the pure rotation spectrum and resort must be made to isotopic substitution in order to determine the molecular parameters. For a non-rigid symmetric top allowance must be made for centrifugal distortion, particularly for the higher rotational levels. Finally, in the same way as for linear molecules, the accuracy of the molecular parameters is limited by the zero-point energy effects.

The molecular parameters of some symmetric tops are given in Table X.

3.4. Asymmetric tops

Molecules with three different moments of inertia are known as *asymmetric tops*. Molecules which fall into this class are those

TABLE X

*Values obtained by microwave spectroscopy of the molecular
parameters of some selected symmetric top molecules*†

Molecule	Bond angle	Bond	r_0	Bond	r_0
CH_3F	HCH 110° 0′	CH	1·109 Å	CF	1·385 Å
CH_3CN	HCH 109° 8′	CH	1·092	CC	1·460
		CN	1·158		
$CHCl_3$	ClCCl 110° 24′	CH	1·073	CCl	1·767
SiH_3Br	HSiH 111° 20′	SiH	1·57	SiBr	2·209
PCl_3	ClPCl 100° 6′	PCl	2·043		
SbH_3	HSbH 91° 30′	SbH	1·712		

with no symmetry, such as CH_3OH (Fig. 1.6), those with one
plane of symmetry, such as CH_2BrCl (Fig. 1.12), those with two
planes of symmetry, such as H_2O (Fig. 1.18), as well as various
other types of molecules with low symmetry. The energy levels
of such molecules cannot be expressed in terms of a single
equation like those of symmetric tops or linear molecules, and
the difference between the rotational energy levels depends on
more than one moment of inertia. Moreover, centrifugal distortion effects are usually rather large. The result is that each
molecule must be treated as an independent problem, and this
becomes very tedious. However, progress has been made in the
interpretation of the very complicated rotation spectra of some
simple asymmetric tops, especially if two of the moments of
inertia have roughly the same value, in which case the spectrum
approximates to that of a symmetric top.

In some cases it has proved possible to determine all three
principal moments of inertia of an asymmetric top. The two
molecular parameters of a bent triatomic molecule, such as H_2O,
of symmetry mm (C_{2v}) may be determined from two of the three
moments of inertia. Since all three moments of inertia of the
water molecule have been determined, it appears at first sight
that we have a redundant independent equation. In fact this is
not so because it can be shown that, for any rigid plane body, the
sum of the two smaller moments of inertia is equal to the third.
Consequently there are only two independent observable

† W. Gordy, W. V. Smith, and R. F. Trambarulo, *Microwave Spectroscopy*.

quantities for a planar molecule. This relation between the moments of inertia can be used to establish the planarity of an asymmetric top. Thus the microwave spectrum of vinylene carbonate (Fig. 5) has been analysed to give the following values for the rotational constants: $B_0^a = 9346 \cdot 70$, $B_0^b = 4188 \cdot 46$, $B_0^c = 2891 \cdot 54$ Mc/s. Using $h = 6 \cdot 62517 \times 10^{-27}$ erg s, we

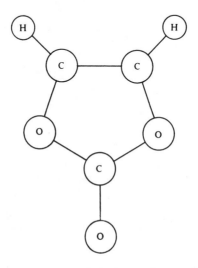

Fig. 5. The molecule of vinylene carbonate, $C_3H_2O_3$.

obtain for the corresponding moments of inertia: $I_0^a = 89 \cdot 773$, $I_0^b = 200 \cdot 334$, $I_0^c \doteq 290 \cdot 188 \times 10^{-40}$ g cm². The sum of I_0^a and I_0^b differs from I_0^c by only $0 \cdot 081 \times 10^{-40}$ g cm², and it may be concluded that the molecule is planar.

The number of simple molecules with dipole moments is limited, and very many of the known ones have now been investigated. Much recent work in the microwave region has been devoted to the investigation of larger asymmetric tops with more molecular parameters than can possibly be obtained even with isotopic substitution. Progress in the elucidation of the molecular parameters of these larger molecules may be made only by assuming values for some of the bond lengths or interbond angles. Examples of the use of this approach are given in Table XI, which lists the molecular parameters of some asymmetric tops.

TABLE XI

Values obtained by microwave spectroscopy of the molecular parameters of some selected asymmetric top molecules†‡

Molecule	Bond angle		Bond	r_0	Bond	r_0
HNCO	HNC	128° 5′	HN	0·987 Å	NC	1·207 Å
	NCO	(180°)	CO	1·171		
CH_2Cl_2	HCH	112° 0′	CH	1·068	CCl	1·772
	ClCCl	111° 47′				
SO_2F_2	OSO	129° 38′	SO	1·370	SF	1·570
	FSF	92° 47′				
O_3	OOO	116° 49′	OO	1·278		
$CH_3SC'N$	HCH	(109°)	CH	(1·09)	CS	1·81
	CSC	142°	C′N	(1·21)	C′S	1·61
CH_3OH	HCH	(109° 28′)	CH	(1·10)	CO	1·421
	COH	110° 15′	OH	0·958		

3.5. Rotational Raman spectra

A molecule need not necessarily possess a permanent dipole moment in order to show a rotational Raman spectrum. Hence homonuclear diatomic molecules may be studied by this method. In any molecule the polarizability in any one direction must be equal to the polarizability in the opposite direction. Consequently, if a diatomic molecule rotates with a frequency ν about an axis at right angles to the internuclear axis, the polarizability must vary with a frequency 2ν. The particular selection rule governing the appearance of rotational Raman lines of a diatomic molecule is $\Delta J = \pm 2$. The energy spacing between two successive levels is, as before,

$$\Delta E_{\rm rot} = B_0 h\{J'(J'+1) - J''(J''+1)\}$$

but with $J' - J'' = 2$. Hence

$$\Delta E_{\rm rot} = B_0 h(4J' - 2),$$

and we obtain the differences shown in Table XII. The spectrum will consist of a series of equally spaced lines on each side of the exciting line, but the frequency separation in this case will be

$$\Delta \nu = 4B_0.$$

† Values in parentheses were assumed in the determination.
‡ W. Gordy, W. V. Smith, and R. F. Trambarulo, *Microwave Spectroscopy*.

Table XII

The pure rotational levels of a rigid rotor in the Raman spectrum

$E_{rot} = B_0\,hJ(J+1)$		$\Delta E_{rot} = B_0\,h(4J'-2)$	J
0			0
$2B_0\,h$	}	$6B_0\,h$	1
$6B_0\,h$	}	$10B_0\,h$	2
$12B_0\,h$	}	$14B_0\,h$	3
$20B_0\,h$	}	$18B_0\,h$	4
$30B_0\,h$	}	$22B_0\,h$	5
$42B_0\,h$	}	$26B_0\,h$	6
$56B_0\,h$			7

Apart from this difference in spacing, the treatment of diatomic molecules and linear polyatomic molecules is identical with our previous treatment of the far infra-red or microwave spectra. A selection of results is given in Table XIII.

Table XIII

Values obtained by rotational Raman spectroscopy of the bond lengths of some selected diatomic and linear molecules

Molecule	r_0
H_2	0·75088 Å
HD	0·74978
D_2	0·74813
F_2	1·418
N_2	1·1001
CS_2	1·5530

The analysis of the rotational Raman spectra of larger molecules is possible only if use is made of the increased resolving power of gratings. Accurate values have been obtained for the molecular parameters of some important symmetric tops whose dimensions were in doubt. The treatment of symmetric tops follows that already given in section 3.3, except that the particular selection rule $\Delta J = \pm 1$ or ± 2, $\Delta K = 0$ applies. Thus two series of lines are obtained, called the R and S branches. If $J'-J'' = 1$, the lines form the R branch with a spacing $\Delta \nu = 2B_0$, and the lines are usually too close to be resolved. If $J'-J'' = 2$, the S branch is formed and the spacing $\Delta \nu = 4B_0$ is twice that in the R branch. From the spacing of the lines in the S branch, the moment of inertia about an axis at right angles to the principal axis can be obtained. Fig. 6 shows the rotational Raman spectra of C_6H_6 and C_6D_6.† In each spectrum the R branch forms an apparent continuum on each side of the exciting line, but a large portion of the S branch can be seen and identified. From an analysis of the S branches, the moments of inertia of the molecules were found to be 147·59 and 178·45 × 10⁻⁴⁰ g cm² respectively. From these moments of inertia the following molecular parameters may be obtained:

$$r_0(\text{CC}) = 1\cdot397 \text{ Å}, \qquad r_0(\text{CH}) = 1\cdot084 \text{ Å}.$$

The method has been applied successfully to about a score of symmetric tops, but is very limited in its scope. If the rotational lines in the S branch are closer than about 0·1 cm⁻¹, they cannot be resolved. This sets an upper limit of about 250 × 10⁻⁴⁰ g cm² on the moment of inertia. Moreover, it is unlikely that the complex spectrum of an asymmetric top will be resolvable.

3.6. Pure rotation spectra of molecules in vibrationally excited states

The pure rotation spectrum of a molecule in a vibrationally excited state may be obtained if the method of detection is sufficiently sensitive. All observations of such spectra have so far been made by microwave spectroscopy and, even then, it is sometimes necessary to heat the sample in order to raise sufficient molecules into the vibrationally excited state. The same gross and particular selection rules apply as to the ground vibrational state, so that we meet nothing new in the interpretation of the spectra. However, the importance of observations taken on excited molecules is that they enable us to obtain values for the equilibrium internuclear distance r_e. For diatomic molecules the moment of inertia will be greater in the excited state than in the ground state, and we may write for the rotational constant of the excited state

$$B_v = B_e - \alpha_e(v + \tfrac{1}{2}),$$

† B. P. Stoicheff, *Can. J. Phys.* 1954, **32**, 339.

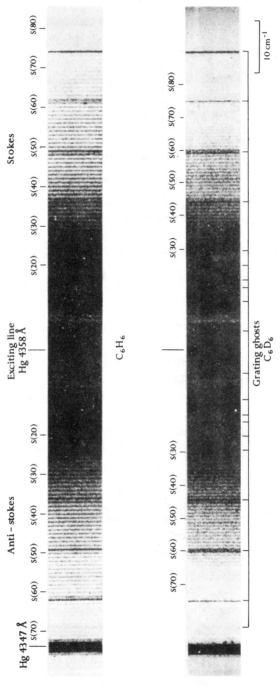

Anti-stokes   Stokes

Hg 4347 Å

Exciting line Hg 4358 Å

s(70) s(60) s(50) s(40) s(30) s(20)   s(20) s(30) s(40) s(50) s(60) s(70) s(80)

C_6H_6

s(70) s(60) s(50) s(40) s(30)   s(30) s(40) s(50) s(60) s(70) s(80)

Grating ghosts
C_6D_6

10 cm^{-1}

Fig. 6. Rotational Raman spectra of C_6H_6 and C_6D_6.

where v is the vibrational quantum number, B_e is the rotational constant in the hypothetical equilibrium position, and α_e is the *rotation-vibration interaction constant*. α_e is usually about one-hundredth part of B_e. In the ground vibrational state, for which $v = 0$,

$$B_0 = B_e - \alpha_e/2.$$

If a value is obtained for the rotational constant in the ground vibrational state and also in any one identified excited state, the magnitudes of B_e and α_e may be determined. From B_e we may find the equilibrium internuclear distance. Some results are shown in Table XIV. In this table the values of r_e and r_0 are quoted to more figures than is warranted by the known accuracy of Planck's constant. We have done this, however, so that we may compare corresponding values for isotopic species.

TABLE XIV

Comparison of r_e and r_0 values of some simple isotopic molecules†

Molecule	B_e	α_e	r_e	r_0
$^{12}C^{16}O$	57897·75 Mc/s	524·16 Mc/s	1·12827 Å	1·13079 Å
$^{13}C^{16}O$	55345·1	488·48	1·12827	1·13072
$^{19}F^{35}Cl$	15483·69	130·67	1·62822	1·63167
$^{19}F^{37}Cl$	15189·22	126·96	1·62821	1·63163

The magnitude and sign of the rotation-vibration interaction constant α_e depend on two opposing effects. The apparent internuclear distance r_0 is given by

$$\frac{1}{r_0} = \left(\frac{1}{r^2}\right)_{av}^{\frac{1}{2}}.$$

The averaging process ensures that $r_0 < r_e$ even if the vibration were harmonic. The effect of anharmonicity, however, tends to make $r_0 > r_e$. For all diatomic molecules so far studied it has been found that α_e is positive and hence $r_0 > r_e$ as shown in Table XIV.

It will be seen from the figures in Table XIV that the r_e values for isotopic species agree extremely well. We have here experimental verification of the validity of the *Born-Oppenheimer*

† W. Gordy, W. V. Smith, and R. F. Trambarulo, *Microwave Spectroscopy*.

approximation which contends that, because of the disparity in mass, the motions of the electrons and the motions of the nuclei can be treated as independent problems with no interaction between them. For our purposes this means that the vibrational and rotational motion can be considered to be quite separate from the electronic motion since the latter, which controls the forces between atoms, involves the charges and configuration of the nuclei but not their masses. This approximation may break down in some special cases, but it is fundamental to the interpretation of molecular spectra since, if it did not apply, we would be unable to set up and solve the requisite wave equations.

Although the r_e values agree so well for different isotopic species, it may be seen from Table XIV that the agreement between the r_0 values is not so good. This is because the rotation-vibration interaction constant depends on the vibration frequency which in turn depends on the atomic masses. The difference between the r_0 values is not large for isotopic diatomic molecules, but becomes larger for polyatomic molecules and, as we saw in section 3.2, restricts the accuracy with which the molecular parameters of polyatomic molecules can be determined by isotopic substitution.

Attempts have been made to obtain r_e values for polyatomic molecules but the situation is complicated by the fact that there are several modes of vibration and each vibration needs its own rotation-vibration interaction constant. The rotational constant of a polyatomic molecule in an excited vibrational state may be written

$$B_{v_1 v_2 \ldots v_n} = B_e - \sum_{i=1}^{n} \alpha_i (v_i + d_i/2), \qquad (11)$$

where d_i is the degeneracy (see section 4.4) of the ith normal mode of vibration. Only for a very few simple linear molecules has it proved possible to obtain values for the different α's, and hence to find the equilibrium internuclear distances.

REFERENCES

G. Herzberg, *Infra-red and Raman Spectra of Polyatomic Molecules* (Van Nostrand).

B. BAK, *Elementary Introduction to Molecular Spectra* (North Holland).

C. H. TOWNES and A. L. SCHAWLOW, *Microwave Spectroscopy* (McGraw-Hill).

W. GORDY, W. V. SMITH, and R. F. TRAMBARULO, *Microwave Spectroscopy* (Wiley).

D. H. WHIFFEN, *Q. Rev. chem. Soc.* 1950, **4**, 131.

IV

VIBRATION-ROTATION SPECTRA

4.1. Diatomic molecules

WE shall begin our consideration of vibration-rotation spectra by a treatment of the ideal case of a diatomic molecule in which the atoms are performing simple harmonic motion. It will be remembered that the gross selection rule demands that the vibration shall produce a change in the dipole moment of the molecule; so that our diatomic molecule must necessarily be heteronuclear. A harmonic oscillator, like a rigid rotor, can be dealt with by wave mechanics. The Schrödinger wave equation can be solved exactly and the vibrational energy levels are given by

$$E_{\text{vib}} = (v + \tfrac{1}{2})h\nu_0, \tag{1}$$

where ν_0 is the *fundamental vibration frequency*, and the vibrational quantum number v can have integral values including zero. The energy levels are equally spaced, as shown in Fig. 2.4, and when $v = 0$, $E_{\text{vib}} = \tfrac{1}{2}h\nu_0$. This half quantum of vibrational energy is the zero-point energy of vibrational motion which was discussed in section 2.3. It is present even at the absolute zero of temperature, and now that we know its magnitude we can express the spectroscopic heat of dissociation D_{e} in terms of the chemical heat of dissociation D_0 by the relation

$$D_{\text{e}} = D_0 + \tfrac{1}{2}h\nu_0.$$

The particular selection rule governing vibrational transitions in a diatomic harmonic oscillator is $\Delta v = \pm 1$, and we might expect that the only frequency to occur in the spectrum will be the fundamental frequency ν_0. In fact this is not so. The pure vibration frequency is scarcely ever seen in the spectrum of a diatomic molecule. The reason for this is that rotation of the molecule is occurring at the same time as vibration but with a frequency that is lower by a factor of 10^2 or 10^3. Suppose we call the frequency of rotation of the molecule ν_{r}. The vibrational and rotational motions will couple together in the same way as the motions of two connected pendulums will couple and, instead of observing the pure vibrational frequency ν_0, we shall see the two frequencies $\nu_0 \pm \nu_r$ lying close to where the pure vibrational frequency should be. Since in practice we are dealing with a large number of molecules which may be rotating with different frequencies and since each molecule emits or absorbs a pair of

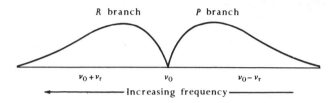

R branch   P branch

$\nu_0 + \nu_r$ ν_0 $\nu_0 - \nu_r$

← ——————— Increasing frequency ———————

FIG. 1. Vibration-rotation band of a heteronuclear diatomic molecule under low resolution.

lines, we shall see a series of lines on each side of the position of the pure vibrational frequency if the lines are resolved. At lower resolution we shall see apparently continuous absorption of the form shown in Fig. 1, the intensity vanishing at the frequency corresponding to the pure vibrational frequency.

4.2. P, Q, and R branches

We can put the conclusions of the previous section into a more exact form. The total vibrational and rotational energy of a diatomic molecule is given by

$$E_{\mathrm{vr}} = (v+\tfrac{1}{2})h\nu_0 + \frac{h^2}{8\pi^2 I} J(J+1).$$

If a transition occurs between two levels of which the upper is denoted by a single prime and the lower by a double prime, we get

$$E'_{\mathrm{vr}} - E''_{\mathrm{vr}} = h\nu_0(v'-v'') + Bh\{J'(J'+1) - J''(J''+1)\},$$

where, as usual, $B = h/8\pi^2 I$. For the moment we shall omit any subscripts on B and I for reasons which will become apparent shortly. For the *fundamental band*, $v'-v'' = 1$. Hence

$$\Delta E_{\mathrm{vr}} = h\nu_0 + Bh\{J'(J'+1) - J''(J''+1)\}. \qquad (2)$$

We must now consider the three possible ways in which the

rotational quantum number can change according to the particular selection rule $\Delta J = 0$ or ± 1. If $J' - J'' = +1$, we may substitute for J'' in equation (2) and we obtain

$$\Delta E_{vr} = h\nu_0 + 2BhJ', \tag{3}$$

where $J' = 1, 2, 3,...$, but not zero, since this would require J'' to be negative. If $J' - J'' = -1$, we may substitute for J' in equation (2) and we get

$$\Delta E_{vr} = h\nu_0 - 2BhJ'', \tag{4}$$

where $J'' = 1, 2, 3,....$. If $J' - J'' = 0$, which, as we shall see, can only occur in special cases,

$$\Delta E_{vr} = h\nu_0. \tag{5}$$

We may combine the three equations (3), (4), and (5) into the form
$$\Delta E_{vr} = h\nu_0 + 2Bhm,$$

where $m = 0, \pm 1, \pm 2, \pm 3,...$.

If $m = 0$, we get a vibrational transition without any accompanying change in the rotational quantum number, and a single strong line called the *zero* or *Q branch* is formed at the origin of the vibration-rotation band. In order to form a Q branch, a molecule must possess angular momentum about the axis joining the nuclei, and the only known case of a stable diatomic molecule showing this type of spectrum in the near infra-red is the odd-electron molecule NO. If m has positive values, a series of equally spaced lines lying on the high-frequency side of the origin is formed, and these lines constitute the *positive* or *R branch*. If m has negative values, the lines form the *negative* or *P branch* on the low-frequency side of the origin. The appearance of such a hypothetical vibration-rotation band is shown in Fig. 2.

In fact it is found that the lines in the P and R branches are not equally spaced and that the Q branch, when it occurs, is not a single line but itself has a fine structure. The reason for this is that the dimensions of the molecule and hence the moments of inertia are not the same in the initial and final stages of the vibrational transition. We may allow for this by writing B'

for the upper state and B'' for the lower. Equation (2) then becomes

$$\Delta E_{\text{vr}} = h\nu_0 + B'hJ'(J'+1) - B''hJ''(J''+1).$$

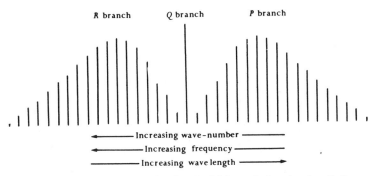

FIG. 2. Vibration-rotation band under high resolution showing P, Q, and R branches.

If we now substitute $J'-J'' = +1$ for the positive or R branch, $J'-J'' = -1$ for the negative or P branch, and $J'-J'' = 0$ for the zero or Q branch, we obtain:

for the R branch,

$$\Delta E_{\text{vr}} = h\nu_0 + h(B'+B'')J' + h(B'-B'')J'^2, \quad J' = 1, 2, 3, \ldots; \quad (6)$$

for the P branch,

$$\Delta E_{\text{vr}} = h\nu_0 - h(B'+B'')J'' + h(B'-B'')J''^2, \quad J'' = 1, 2, 3, \ldots; \quad (7)$$

for the Q branch,

$$\Delta E_{\text{vr}} = h\nu_0 + h(B'-B'')J' + h(B'-B'')J'^2, \quad J' = 0, 1, 2, \ldots. \quad (8)$$

In a diatomic molecule the moment of inertia is always greater in a vibrationally excited state than it is in the ground state, i.e. $I' > I''$ and $B' < B''$. In this case the last term in the three equations (6), (7), and (8) is negative. In the R branch this term will gradually outweigh the second term as J' increases and cause ΔE_{vr} to start decreasing again after an initial increase. The R branch is said to show a *head* where the lines bunch together in a characteristic and easily recognizable fashion. The head lies on the high frequency side of the origin and the branch *degrades to the red.* The P and Q branches will show lines whose frequency steadily diminishes as the rotational quantum number increases.

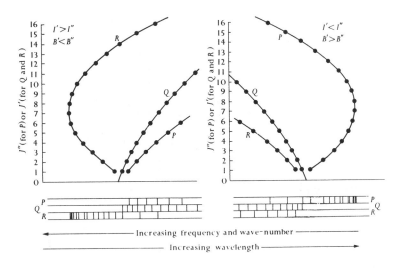

FIG. 3. Relation between P, Q, and R branches shown in the form of a Fortrat diagram.

In linear polyatomic molecules (for which the above equations are also applicable), and in diatomic molecules when an electronic transition is involved, the moment of inertia in an excited state is not necessarily greater than in a lower state, and we must allow for the possibility that $I' < I''$ and $B' > B''$. In this case the P branch shows the head, but *degrades to the violet,* whereas the Q and R branches show lines with steadily increasing frequency. These relationships are illustrated by the so-called *Fortrat diagrams* in Fig. 3.

In the vibration-rotation spectra of simple molecules it is unusual for a branch to show a head as the difference in the moments of inertia is too small. We would have to go to very high values of the rotational quantum number for the last term in equations (6) or (7) to outweigh the term in J' or J''. The spacing of the lines is not constant, however, and from the positions of the lines it is possible, as we shall see in section 4.5, to evaluate the moments of inertia in the two vibrational states. If a branch does show a head, whether in the vibration-rotation or in the electronic spectrum, it is clear that, by noting whether the branch degrades to the red or to the violet, we can tell immediately if the moment of inertia is greater in the excited state or if it is greater in the lower state.

4.3. Anharmonicity

If we wish to deal with real molecules, we must abandon our approximation of a harmonic oscillator and we then know (section 2.3) that the vibrational levels of the molecule become closer and closer together as we approach the dissociation limit. We may allow for the effect of anharmonicity by modifying equation (1) for the vibrational energy levels of a harmonic oscillator and writing

$$E_{vib} = (v + \tfrac{1}{2})h\nu_0 - (v + \tfrac{1}{2})^2 x h\nu_0,$$

where x is the *anharmonicity constant*. Anharmonicity has some important consequences. In the first place, as we have seen, the spacing between the vibrational energy levels becomes smaller as the vibrational quantum number increases and, if we can measure the progressive diminution in spacing, we can obtain a value for x and from this a value for the spectroscopic heat of dissociation D_e. Secondly, the particular selection rule $\Delta v = \pm 1$ is no longer valid and the vibrational quantum can change by $\pm 1, \pm 2, \pm 3, \ldots$. The spectrum will therefore consist of *harmonics* or *overtones*, as well as the fundamental band. Finally, in polyatomic molecules, *combination bands* or *difference bands*, involving two or more modes of vibration, may occur. The presence of these additional bands results in a complex vibration-rotation spectrum even for triatomic molecules.

4.4. Normal modes

We may regard a polyatomic molecule as a number of mass points joined together by springs. If such a system is struck, it will perform a complicated vibrational motion, known as a

Lissajous motion, which, in general, never repeats itself. It is found, however, that even the most complicated molecular vibration can be resolved into a relatively small number of *normal vibrations* or *normal modes*. Each normal mode can be described by a set of *normal coordinates* which tell us how the atoms are vibrating. Each normal mode is independent in the sense that, if a molecule happens to vibrate for a long time in a single normal mode—a thing that is impossible in practice because of molecular collisions—no other vibration will be excited. In any one normal mode all the nuclei vibrate in phase and with the same frequency. We may obtain the number or normal modes by the following simple consideration.

Each of the N atoms in a non-linear polyatomic molecule has three degrees of freedom corresponding to motion along the three Cartesian coordinates. The whole molecule has, therefore, $3N$ degrees of freedom. However, three of these correspond to translation of the molecule along the three axes, and three correspond to an overall rotation of the molecule about the three axes. This leaves $3N-6$ normal modes of vibration. In a linear molecule there is one more mode of vibration giving a total of $3N-5$. The reason for this extra mode of vibration can be seen from Fig. 4. Fig. 4 (*a*) shows a bent triatomic molecule AB_2. Two of the atoms are moving upwards and one downwards, resulting in a rotation of the molecule round an axis in the plane of the paper. It can be seen from Fig. 4 (*b*) that the same movements of the atoms of a linear molecule correspond not to a rotation but to a *bending vibration* in which the BAB angle is altered. A diatomic molecule has $3 \times 2 - 5 = 1$ normal mode, in which the internuclear distance successively increases and decreases. This is known as a *stretching vibration*.

A pentatomic molecule, such as CH_4 (Fig. 1.21), should have $3 \times 5 - 6 = 9$ normal modes and hence nine vibrational frequencies. In fact CH_4 is found to have only four different fundamental frequencies. The reason for this is that there are really still nine frequencies but some of them, by virtue of the symmetry of the molecule, have identical values. When two or three frequencies have the same magnitude we say that the vibrations are *doubly-*

or *triply-degenerate* respectively. CH_4 has two triply-degenerate vibrations, one doubly-degenerate vibration, and one non-degenerate vibration. A molecule necessarily has degenerate vibrations if it possesses a pure rotation axis of order three or more. All symmetric tops, therefore, have degenerate vibrations.

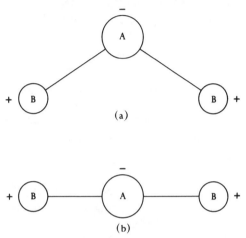

FIG. 4. Comparable nuclear motions in triatomic molecules: (a) bent molecule with motions leading to rotation of the molecule, (b) linear molecule with motions leading to a bending vibration.

Triple degeneracy can occur only if the molecule possesses more than one threefold axis, and is thus restricted to tetrahedral or octahedral molecules (Table I).

It is clear that, if symmetry restricts the number of distinct vibrations in this way, we may reverse the process and, by counting the number of distinct vibrations, obtain information about the molecular symmetry. We shall see later that a simultaneous investigation of the vibrational Raman spectrum can increase the certainty of the information obtained in this way, but, even so, considerable care is necessary. There are several reasons why it may not be possible to count all the fundamental frequencies: some may be too weak to be observed, or lie outside the range of observation, or two frequencies may by chance have the same or nearly the same value, and this *accidental degeneracy* can be very misleading.

The number of normal modes that a molecule possesses depends, as we have seen, on the number of atoms. The degeneracy depends on the point group to which the molecule belongs. For instance, all bent triatomic molecules of the general formula AB_2 (e.g. SO_2) belong to the point group mm (C_{2v}) and have three non-degenerate vibrations. All tetrahedral molecules of the general formula XAB_3 (e.g. CH_3Cl) belong to the point group $3m$ (C_{3v}) and have three doubly-degenerate and three non-degenerate vibrations. The vibrations will be active in the infra-red only if they involve a change in the dipole moment of the molecule, and active in the Raman only if they involve a change in polarizability of the molecule. It is possible for some vibrations to involve neither a change in dipole moment nor in polarizability, and to be inactive in both the infra-red and in the Raman.

The number, the degeneracy, and the spectral activity of the normal modes of molecules which have many of the probable combinations of point symmetry and number of atoms have been evaluated mathematically and the results tabulated. If the formula of the molecule is known and if the number of fundamentals active in the infra-red and Raman can be reliably determined, it is possible to refer to these tables and to decide to which point groups the molecule might belong. We shall consider in detail an example of such a determination of molecular symmetry in section 4.7.

4.5. The vibration-rotation spectrum of CO_2

The linear CO_2 molecule of ∞/mm ($D_{\infty h}$) symmetry possesses four normal modes of vibration. These are illustrated in Fig. 5. The symmetric stretching vibration ν_1 does not cause a change in the dipole moment of the molecule, since the molecule has zero dipole moment when it is compressed, when it is extended, and in all intermediate positions. ν_1 will, therefore, be inactive in the infra-red. Vibrations which are symmetric with respect to every symmetry element in the molecule are known as *totally symmetric vibrations* and are never active in the infra-red. ν_2 is the bending vibration and, since the atoms can perform similar motions in two independent directions at right angles to each other, the vibration is doubly-degenerate. When CO_2 is bent the molecule will possess a dipole moment, so that, during the bending vibration, the dipole moment will change from a positive finite value, through zero, to a negative finite value.

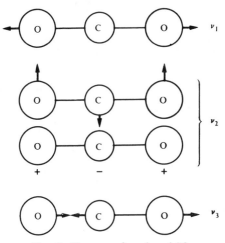

FIG. 5. The normal modes of CO_2.

Since the dipole moment changes in this way, the molecule can interact with incident electromagnetic radiation and ν_2 will be active in the infra-red. ν_3 is an asymmetric stretching vibration. Clearly, when CO_2 vibrates in accordance with ν_3, it will possess a finite dipole moment of opposite sign at each end of the motion. Thus ν_3 will also be active in the infra-red.

In order to deduce the Raman activity, we have to consider the change in polarizability instead of the change in dipole moment. When CO_2 is vibrating in accordance with ν_1, the polarizability will change from a small value when the molecule is compressed, through the normal value at the equilibrium configuration, to a high value when the molecule is stretched. ν_1 will, therefore, be Raman active. On the other hand, for both ν_2 and ν_3 the polarizability is identical at opposite ends of the vibration, since we cannot attach a sign to the polarizability as we can to the dipole moment. Consequently both ν_2 and ν_3

are inactive in the Raman. We conclude, therefore, that, in the infra-red, ν_1 is inactive and ν_2 and ν_3 are active, whereas in the Raman, ν_1 is active and ν_2 and ν_3 are inactive. We have here an illustration of an important rule known as the *rule of mutual exclusion*. If a molecule possesses a centre of symmetry, no fundamental which is active in the infra-red can be active in the Raman, and no fundamental that is active in the Raman can be active in the infra-red.

More than forty different bands have been observed between 1 and 15 μ in the infra-red and Raman spectra of CO_2. The following values have been assigned to the three fundamentals:

$$\nu_1 \cong 1{,}340 \text{ cm}^{-1}, \quad \nu_2 = 667 \text{ cm}^{-1}, \quad \nu_3 = 2{,}349 \text{ cm}^{-1}.$$

It is an important part of a spectroscopist's task to assign frequencies to the various normal modes. This is done by a consideration of a number of factors. In the first place, we know how many fundamentals to expect, and whether they will be found in the infra-red spectrum, in the Raman spectrum, in both, or in neither. Secondly, experience has shown that fundamentals are generally much more intense than overtone, combination, or difference bands. Thirdly, we know also that bending vibrations usually have lower frequencies than stretching vibrations. Finally, the appearance of the rotational fine structure or, if the fine structure is not resolved, of the *band contour* can be of great assistance. In the vibration ν_3 the changing dipole moment is always parallel to the principal axis of rotation of the molecule, which in this case is the internuclear axis. Such a vibration forms what is known as a *parallel band*, which is very similar to the fundamental band of a diatomic molecule. A change in the vibrational quantum number cannot occur without a simultaneous change in the rotational quantum number, and there is consequently no Q branch. In the ν_2 vibration there is a perpendicular component to the dipole formed during the motion of the nuclei, so that ν_2 forms a *perpendicular band*. The particular selection rule is now $\Delta J = 0$ or ± 1, and a perpendicular band will therefore have a Q branch. Fig. 6 shows the fine structure of the ν_2 band of CO_2 at 15 μ and the central Q branch is clearly visible.† Fig. 7 shows the ν_3 band at 4·3 μ and the absence of a Q branch shows the vibration to be of the parallel type.†

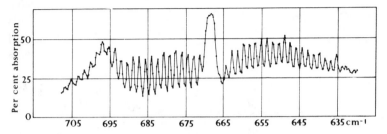

FIG. 6. The ν_2 band of CO_2. The band is of the perpendicular type with a central Q branch.

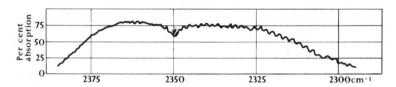

FIG. 7. The ν_3 band of CO_2. The band is of the parallel type with no Q branch.

Actually the spectrum of CO_2 is a good deal more complicated than the previous discussion has indicated. In the first place the Raman line at ~ 1340 cm^{-1} is found to consist of two lines with frequencies of 1286 and 1388 cm^{-1}. The occurrence of this doubling lies in the fact that the first harmonic of the ν_2 vibration is almost equal in frequency to the ν_1 vibration, i.e. $2\nu_2 \cong \nu_1$. Consequently a perturbation of the levels, known as *Fermi resonance*, takes place, the levels splitting into two components of roughly equal intensity. Fermi resonances are fairly common and they introduce what is apparently an additional and unexpected fundamental into the spectrum. Another feature of the vibration-rotation spectrum of CO_2 is seen in the rotational fine structure. The P and R branches of both ν_2 and ν_3 can be resolved under high resolution, and the spacing of the

† P. E. Martin and E. F. Barker, *Phys. Rev.* 1932, **41**, 291.

rotational lines is found to be the same in both fundamentals, confirming that the molecule has only one moment of inertia and that CO_2 is linear. However, the spacing between successive rotational lines is twice the expected value. The reason for this is that alternate rotational levels of molecules of ∞/mm $(D_{\infty h})$ symmetry have different statistical weights. This, of course, is

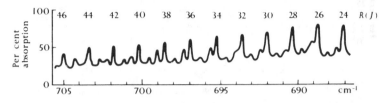

Fig. 8. Part of the R branch of the ν_2 band of CO_2 under high resolution.

the reason for the existence of the ortho and para forms of hydrogen. The statistical weights depend on the spin of the nuclei and, in a special case like CO_2 where the oxygen nuclei have zero spin, alternate levels have zero statistical weight. Thus alternate lines are completely absent, only those lines with even values of J appearing. This is illustrated in Fig. 8 which shows part of the R branch of the ν_2 vibration.† This alternating intensity effect can be of use in the determination of the symmetry of linear molecules since, for molecules of ∞m $(C_{\infty v})$ symmetry, all the rotational levels have the same statistical weight and there should be no intensity alternation. It was shown in this way that the N_2O molecule must be N–N–O and not N–O–N.

If the observed lines in the R and P branches are fitted to equations (6) and (7), it is possible to obtain values for $B' + B''$ and $B' - B''$, and hence to find B' and B'', the rotational constants in the upper and lower vibrational states. Actually this proves to be a rather inaccurate way of finding the rotational constants, and an analytical procedure known as the *method of combination differences* is usually used instead. Suppose, in the P branch, we

† K. Rossman, K. N. Rao, and H. H. Nielsen, *J. chem. Phys.* 1956, **24**, 103.

select a line which has a given value of J'' for the lower rotational quantum number, and suppose that we pick out in the R branch the line that is associated with this same value of J''. Then from equations (6) and (7), and remembering that $J' = J''+1$, we get for the energy difference between these lines

$$R(J'')-P(J'') = h(B'+B'')(J''+1)+h(B'-B'')(J''+1)^2+$$
$$+h(B'+B'')J''-h(B'-B'')J''^2$$
$$= 4hB'(J''+\tfrac{1}{2}).$$

Similarly, if we select the next higher line in the P branch and the next lower line in the R branch, we get for the energy difference

$$R(J''-1)-P(J''+1) = h(B'+B'')J''+h(B'-B'')J''^2+$$
$$+h(B'+B'')(J''+1)-h(B'-B'')(J''+1)^2$$
$$= 4hB''(J''+\tfrac{1}{2}).$$

In other words, by appropriate selection of the lines, we can eliminate either B'' or B'. If we now plot the observed energy differences against J'', we should obtain a straight line whose slope gives us directly $4hB'$ in the first case and $4hB''$ in the second. All the graphs plotted in this way tend to deviate from straight lines at high values of J''. This is because we have not taken into consideration in the above treatment the centrifugal distortion constant D. From the deviations from linearity it is possible to obtain values for this constant in the upper and lower vibrational states.

We can obtain the difference between the rotational constants in the upper and lower states with even greater accuracy if we make use of the combination sum

$$R(J''-1)+P(J'') = 2h\nu_0+2h(B'-B'')J''^2.$$

A plot of the left-hand side of this equation against J''^2 gives us a straight line whose slope is $2h(B'-B'')$ and whose intercept gives an accurate value for ν_0, the fundamental vibration frequency.

From an analysis of the ν_2 vibration of CO_2 carried out in this way, it was found that B_{000} was equal to 0.39016 cm^{-1} and that

$B_{010} - B_{000}$ (i.e. the difference between the rotational constants in the state for which $v_1 = 0, v_2 = 1$, and $v_3 = 0$ and the ground state in which all three vibrational quantum numbers are zero) was 0.00041 cm^{-1}. Since the rotational constant in the excited state is greater than in the ground state, the moment of inertia of CO_2 in the ground state is greater than in the excited state. This is usually the case for perpendicular vibrations of a linear polyatomic molecule.

There is only one molecular parameter in a molecule of ∞/mm $(D_{\infty h})$ symmetry, so that we may use the values of B_{000} and B_{010} given above to determine the C—O bond length in the ground state and in the state in which v_2 has been excited by the addition of one quantum. Using the constants given in the appendix, and taking $^{16}O = 16.0000$, we find $r_0(CO) = 1.1622$ Å for the ground state and $r_0(CO) = 1.1616$ Å for the excited state.

4.6. Vibration-rotation spectra of non-linear molecules

The spectra of non-linear molecules are much more complex than those of linear ones. It is seldom that a simple pictorial derivation of the normal modes can be made as we have done for CO_2, and it is necessary to resort to group theory in order to elucidate the nature of the normal vibrations. We shall not discuss the mathematical derivation, but merely give some indications of how the frequencies may be assigned to the normal modes.

Reasonably satisfactory frequency assignments have been made for most stable molecules which possess some symmetry and which have less than about twelve atoms. For symmetric tops we can still classify the vibrations as parallel and perpendicular to the main rotation axis of the molecule. Parallel vibrations of a symmetric top give bands which resemble perpendicular bands of linear molecules and have a central Q branch. Perpendicular bands of a symmetric top can be distinguished from parallel bands since the former are seen to consist of a number of sub-bands each with P, Q, and R branches. It is not possible to classify the vibrations of an asymmetric top as parallel and perpendicular, and we speak of type A, B, or C

bands, depending on whether the change of dipole moment is in the direction of the least, intermediate, or largest moment of inertia respectively. The three types of bands have characteristic differences in shape which help in the assignment.

Much assistance may be gained from the vibrational Raman spectrum since it is usually far simpler than the infra-red. The fundamental frequencies, provided there are no Fermi resonances, are much more intense than overtone or combination bands. Moreover, the most intense fundamentals are usually due to symmetric vibrations. Observations on the state of polarization of the Raman lines also help. Suppose we label the direction of the incident light as the z-axis. If we now perform two experiments, one with the incident light polarized parallel to the xz-plane and the other with the light polarized perpendicular to this plane, we find that some lines may have the same intensity in the two experiments whereas others may be of very different intensity. The lines with different intensity are polarized and correspond to totally symmetric vibrations. The lines with the same intensity are said to be *depolarized* and correspond to vibrations which are not totally symmetric.

If a frequency is inactive in the Raman and in the infra-red, it may sometimes be obtained from a combination or a difference band, but it often has to be obtained by indirect methods from the observed values of such thermodynamic properties as the heat capacity or the entropy, although indirect methods, since they depend on a small difference between two large quantities, yield only crude values for an inactive frequency. Finally, if the molecule under investigation contains hydrogen, a study of the deuterium analogues can be of great assistance in sorting out those frequencies which depend on the motion of the hydrogen atom, since X—H stretching frequencies lie in the range 3000 ± 300 cm^{-1}, whereas X—D frequencies are about 500 cm^{-1} less. If, then, the spectrum of a molecule is compared with that of its perdeutero analogue, some frequencies will move to lower frequencies, and these will belong to motions that involve mainly the hydrogen or deuterium atoms.

It will be seen that a full frequency assignment is a formidable task. Evidence from many different sources must be sifted before a final decision is made. Often additional evidence is then found which necessitates a change in the assignment, and even such apparently simple molecules as ethylene C_2H_4 are constantly under review.

For asymmetric molecules with more than about six atoms, and for any molecule with more than about twelve atoms, no full assignment can be made. However, a study of the complex vibrational infra-red spectrum, even under low resolution, is extremely useful for analytical and diagnostic purposes. The infra-red spectrum of a compound is characteristic of that compound, and the spectrum of a mixture is the superposition of the spectra of the individual constituents. It is found, moreover, that certain groups of atoms always absorb at roughly the same frequency, no matter in what molecule they are located. For instance C—H stretching frequencies lie in the range 2700–3300 cm^{-1}; organic nitriles show strong absorption between 2200 and 2400 cm^{-1}; ionic sulphates have two regions of absorption, one at about 450 cm^{-1} and the other at about 1100 cm^{-1}. Such observations enable us to detect the presence of certain groups of atoms in the most complicated molecules.

4.7. The determination of molecular symmetry

We may illustrate a typical application of the combined use of infra-red and Raman spectroscopy for the determination of molecular symmetry by a consideration of SF_4.† Most AX_4 molecules are regular tetrahedra, but the valency shell of the sulphur atom in SX_4 molecules has ten electrons, and there are reasons for believing that the extra pair of inert electrons might lower the symmetry. If SF_4 has $\overline{4}3m$ (T_d) symmetry, four distinct fundamental frequencies should be seen, as in CH_4. Two should be infra-red active and all four Raman active. Of the Raman lines one should correspond to a totally symmetric vibration and should be polarized. If the molecule is essentially a trigonal bipyramid with one of the equatorial atoms removed (Fig. 9), it

† R. E. Dodd, L. A. Woodward, and H. L. Roberts, *Trans. Faraday Soc.* 1956, **52**, 1052.

will have *mm* (C_{2v}) symmetry. In this case there should be nine distinct fundamentals of which eight should be infra-red active and all nine Raman active. Four of the Raman lines should be polarized. Alternatively one of the polar atoms of the trigonal bipyramid may be absent (Fig. 10) and the molecule will then have *3m* (C_{3v}) symmetry. There should now be six distinct fundamentals all of which are infra-red and Raman active. Three of the Raman lines should be polarized. These three possible point

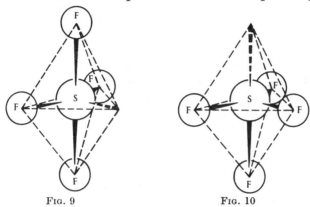

FIG. 9.   FIG. 10.

FIG. 9.   Possible shape of the molecule of sulphur tetrafluoride, SF_4.
A trigonal bipyramid with one equatorial atom removed.
FIG. 10.   Possible shape of the molecule of sulphur tetrafluoride, SF_4.
A trigonal bipyramid with one polar atom removed.

groups with the possible spectral activity are summarized in Table XV. The infra-red spectrum of the gas at room temperature was recorded for the range 400–5000 cm^{-1}. The Raman spectrum of the liquid at $-60°$ C was obtained photographically. The incident radiation from a mercury-arc was filtered to allow

TABLE XV

Possible spectral activity of a pentatomic molecule of unknown symmetry

Symmetry	Fundamentals	Infra-red	Raman	Polarized lines
mm (C_{2v})	9	8	9	4
$3m$ (C_{3v})	6	6	6	3
$\bar{4}3m$ (T_d)	4	2	4	1

only a wavelength of 4358 Å through. Raman shifts in the range 100–1000 cm^{-1} could be determined.

At least five fundamentals could be detected in the infra-red and at least five in the Raman, one of which was polarized. This evidence is sufficient to exclude the possibility that SF_4 is a regular tetrahedron, but it cannot distinguish between mm (C_{2v}) and $3m$ (C_{3v}) symmetry. The failure to observe some of the fundamentals may be due to a number of reasons. They may be too weak; they may lie outside the range of observation; they may be obscured by a combination or a difference band; or two fundamentals may accidentally have about the same frequency. A study of the shape of the infra-red bands does make it probable that the molecule has mm (C_{2v}) rather than $3m$ (C_{3v}) symmetry, and this conclusion has now been confirmed by other methods, but it cannot be regarded as having been conclusively established from infra-red and Raman spectra alone.

4.8. The vibration-rotation spectrum of C_2HD

The infra-red and Raman spectra of acetylene, deuteroacetylene, and monodeuteroacetylene have been the subjects of a great deal of study. We shall consider C_2HD in some detail since it illustrates most of the points mentioned in this chapter and the previous one. Early work on C_2H_2 showed that the molecule had only one moment of inertia and hence that the molecule was linear. The presence of the alternating intensity effect, with lines of odd J three times as intense as those with even J, proved the presence of a centre of symmetry. C_2H_2 thus has ∞/mm ($D_{\infty h}$) symmetry. We shall, therefore, assume that C_2HD is linear but it cannot, of course, possess a centre of symmetry. C_2HD has $3 \times 4 - 5 = 7$ normal modes which are shown in Fig. 11. The last two are doubly-degenerate bending vibrations, whereas the first three are non-degenerate stretching vibrations. This gives a total of five distinct fundamentals all of which should be active in both the Raman and in the infra-red. Several workers have observed all five fundamentals in the infra-red. Less work has been done on the Raman spectrum and no observations have been made on the state of polarization of the Raman lines. Only two Raman lines have been observed, but in some ways this is an advantage for our purposes because the most intense lines in the Raman spectrum are probably due to totally symmetric vibrations.

Table XVI shows the observed infra-red and Raman frequencies of C_2HD with details of the intensity and structure of

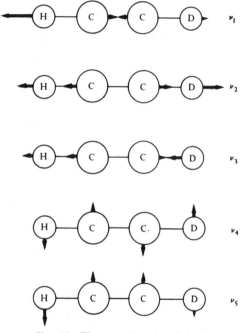

FIG. 11. The normal modes of C_2HD.

the various bands and lines. The intense band at 3334·8 cm^{-1} in the infra-red is also observed in the Raman spectrum. The fact that it is observed at all in the Raman suggests that it is probably due to a totally symmetric vibration. The frequency is in the range expected for a C—H stretching vibration. The infra-red band is of the parallel type with no central Q branch. These three pieces of evidence strongly suggest that the frequency should be assigned to ν_1 which involves almost entirely a stretching of the C—H bond. The vibration ν_3 involves mainly a

TABLE XVI

The observed vibration spectrum of monodeuteroacetylene†

ν_{obs}	Band type‡			Assignment
518·8 cm^{-1}	I. Perp.	S.		ν_4
683	I. Perp.	S.		ν_5
1034·0	I. Par.	M.		$2\nu_4$
1200·3	I. Par.	M.		$\nu_4 + \nu_5$
1330	I. Perp.	W.		$\nu_2 - \nu_4$
1338·3	I. Par.	M.		$2\nu_5$
1853·8	{ I. Par.	M.		ν_2
	{ R.	S.		
2011	I. Perp.	M.		$3\nu_5$
2065	I. Perp.	W.		$\nu_3 - \nu_4$
2583·6	I. Par.	S.		ν_3
3334·8	{ I. Par.	S.		ν_1
	{ R.	S.		
3950	I.	W.		{ $\nu_1 + \nu_4$
				{ $\nu_1 + \nu_5$
5100	I.	W.		{ $2\nu_3$
				{ $\nu_1 + \nu_2$
8409·4	I. Par.	M.		$2\nu_1 + \nu_2$
9050·6	I. Perp.	W.		$2\nu_1 + \nu_2 + \nu_5$
9138·9	I. Par.	M.		$2\nu_1 + \nu_3$
9404·8	I. Par.	M.		$\nu_2 + 3\nu_3$
9706·4	I. Par.	S.		$3\nu_1$

‡ I. = infra-red.   R. = Raman.   Perp. = perpendicular type band.
Par. = parallel type band.   S. = strong.   M. = medium.   W. = weak.

stretching of the C—D bond, but it is an antisymmetric stretching vibration. It should thus appear with reduced intensity in the Raman spectrum but should appear as a parallel type band in the infra-red with a frequency of about 2500 cm^{-1}. The band at 2583·6 cm^{-1} satisfies these criteria and may be assigned to ν_3. The vibration ν_2 is again a totally symmetric stretching vibration, but it involves largely a stretching of the C≡C bond. Consequently we should expect another parallel-type band of rather lower frequency, appearing as an intense line also in the Raman. The band at 1855·8 cm^{-1} is the obvious choice and is assigned to ν_2. Two strong perpendicular-type bands with central Q branches are found in the infra-red, and their frequencies are in the right range for them to be assigned to the two bending

† Based on G. Herzberg, *Infra-red and Raman Spectra of Polyatomic Molecules.*

modes. No decision can be reached from the spectra as to which frequency should be assigned to ν_4 and which to ν_5, but theoretical considerations lead to the conclusion that bending of the type ν_4 should be easier and thus have a lower frequency than that of ν_5. Consequently the band at 518·8 cm^{-1} is assigned to ν_4 and 683 cm^{-1} to ν_5, although the reverse assignment was originally made. A complete assignment of all the fundamental, overtone, combination, and difference bands so far observed between 1 and 20 μ is given in the last column of Table XVI.

A thorough investigation has been made at high resolution of a number of the fundamental and combination bands of C_2HD. The ν_3 band, partially overlaid by two weaker bands, is shown in Fig. 12.† The rotational constant in the ground state B_{00000} was found from the ν_3 band by the method of combination differences (section 4.5) and has the value 0·99105 cm^{-1}. The rotational constants B_v and the centrifugal distortion constants D_v in some of the vibrationally excited states were also determined by the same method. The five rotation-vibration interaction constants were found by the method described in section 3.6. As usual α was found to be positive for the stretching vibrations and negative for the bending vibrations as shown in Table XVII. From the values for B_{00000} and the α's, B_e could be determined from equation (3.11), and a value of 0·9948 cm^{-1} was obtained. A similar study of C_2D_2 had already yielded $B_{00000} = 0·8479$ cm^{-1} and $B_e = 0·8503$ cm^{-1}. The rotational constants of C_2HD and C_2D_2 were then combined as described in section 3.2. The resulting apparent and equilibrium internuclear distances are shown in Table XVIII.

TABLE XVII

The rotation-vibration interaction constants of monodeuteroacetylene

α_1 +0·00499 cm^{-1}
α_2 +0·00439
α_3 +0·00678
α_4 −0·00322
α_5 −0·0011

† J. Overend and H. W. Thompson, *Proc. R. Soc.* A 1956, **234**, 306.

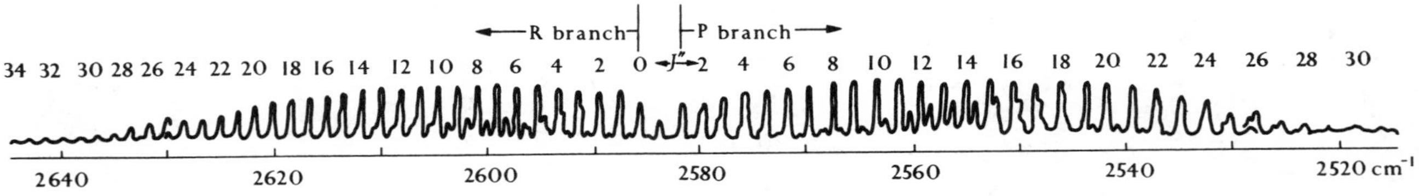

Fig. 12. The ν_3 band of C_2HD under high resolution.

TABLE XVIII

Apparent and equilibrium internuclear distances
in the deuteroacetylenes

	$r(CH)$	$r(CC)$
r_0	1·0537 Å	1·2105 Å
r_e	1·0587	1·2050

REFERENCES

G. HERZBERG, *Infra-red and Raman Spectra of Polyatomic Molecules* (Van Nostrand).

B. BAK, *Elementary Introduction to Molecular Spectra* (North Holland).

L. A. WOODWARD, *Q. Rev. chem. Soc.* 1956, **10**, 185.

PART II

Diffraction Methods

V

ELECTRON DIFFRACTION

5.1. The scattering of electrons by gases

IF a beam of electrons is accelerated through about 40,000 volts and passed through a sample of gas, diffraction effects will occur since the wavelength of the electrons is of the order of magnitude of internuclear distances in molecules. The wavelength of the electrons is given by the de Broglie relation

$$\lambda = \frac{h}{m_e v}, \tag{1}$$

where m_e is the mass of an electron and v the velocity. If the electrons are accelerated through a potential difference V, their kinetic energy is given by

$$\tfrac{1}{2} m_e v^2 = eV, \tag{2}$$

where e is the electronic charge. We may eliminate v from equations (1) and (2) and obtain for the wavelength of the electrons

$$\lambda = \frac{h}{m_e} \bigg/ \sqrt{\left(\frac{m_e}{2eV}\right)}.$$

For $V = 40$ kilovolts this gives a wavelength of about 0·06 Å. For accurate work a relativity correction has to be applied to the wavelength, since the electrons are moving with a velocity which is not negligible compared with the velocity of light.

Most electron diffraction work has been confined to gases, but some very interesting experiments have been carried out with solid samples. Since electrons can penetrate only a thin layer of a solid, electron diffraction is suitable for the study of surface effects. A few molecular structure determinations have been performed with very thin single crystals, but we shall restrict our discussion to electron diffraction by gases.

In a typical experiment a beam of electrons from a hot filament is monochromatized by means of a carefully stabilized accelerating potential. The beam traverses a magnetic focusing system and then passes through a stream of the sample gas. The gas is introduced momentarily into the system through a nozzle and is immediately condensed on to a cold trap on the far side of the electron beam. After diffraction the electrons strike a photographic plate which is exposed synchronously with the introduction of the sample gas for a second or less. The whole apparatus must be evacuated to a pressure of about 10^{-5} mm of Hg, otherwise the electrons will all be absorbed before they reach the photographic plate. Any stable substance that will give a vapour pressure of a few mm of Hg at reasonable temperatures may be used as the sample.

The electrons in the beam are scattered by the potential field of the sample molecules. The diffraction of electrons is much greater than that of X-rays (section 8.1) because the electrons carry a negative charge, but both electrons and X-rays are scattered more by heavy atoms than by light ones. The molecules in the stream of the sample gas are randomly oriented so that the diffraction pattern that is produced must be radially symmetric. The total diffraction pattern I_T consists of three parts, the *incoherent atomic scattering* I_i, the *coherent atomic scattering* I_a, and the *coherent molecular scattering* I_m. Thus

$$I_T = I_i + I_a + I_m.$$

The incoherent atomic scattering arises because an incident electron may raise an atom to any one of a number of higher energy levels. The electron loses a corresponding amount of energy and consequently the electrons that suffer incoherent scattering will have a whole spectrum of wavelengths. The incoherent atomic scattering falls off very rapidly with θ, the angle between the incident and the scattered beam, and shows no maxima or minima. The intensity of the coherent scattering $I_c = I_a + I_m$ is given by

$$I_c = k \left\{ \sum_{j=1}^{N} \frac{F_j^2}{s^4} + \sum_{\substack{j=1 \\ (j \neq k)}}^{N} \sum_{k=1}^{N} \frac{F_j F_k}{s^4} A_{jk} \frac{\sin(sr_{jk})}{sr_{jk}} \right\}. \qquad (3)$$

In this equation k is a constant which depends on the geometry of the apparatus, r_{jk} is the apparent distance between the two atoms j and k (which may or may not be joined by a chemical bond), and A_{jk} is an exponential term which allows for the fact that the atoms within the N-atomic molecule are not strictly at rest but are vibrating with respect to each other. A_{jk} has the form $\exp(-\langle \frac{1}{2} l_{jk}^2 \rangle_{av} s^2)$, where $\langle l_{jk}^2 \rangle_{av}$ is the mean square amplitude of vibration between the jth and kth atoms. s is a parameter which depends on the scattering angle θ and the wavelength of the electrons λ, and is given by $s = (4\pi \sin \theta/2)/\lambda$. F_j and F_k are the *scattering factors for electrons* of the two atoms j and k. F_i is equal to $Z_i - f_i$, where Z_i is the atomic number of the ith atom and f_i is the atomic scattering factor for X-rays (section 8.1). It will be seen that the scattering factor F_i contains a positive term Z_i which is due to the positively charged nucleus, and a negative term f_i which arises from the screening effect of the extra-nuclear electron cloud.

The first term on the right-hand side of equation (3) is the coherent atomic scattering which, like the incoherent atomic scattering, is monotonic and falls off rapidly with θ. The second term is the molecular scattering and is the term in which we are interested. Since the molecular scattering depends on a term of the general form $\sin x/x$, and since x depends on θ, the intensity of the molecular scattering

$$I_m = k \left\{ \sum_{\substack{j=1 \\ (j \neq k)}}^{N} \sum_{k=1}^{N} \frac{F_j F_k}{s^4} A_{jk} \frac{\sin(sr_{jk})}{sr_{jk}} \right\} \qquad (4)$$

must pass through a series of maxima and minima as θ increases. Since the total scattering consists of the coherent molecular scattering superimposed on the coherent and the incoherent atomic scattering, the total diffraction pattern will appear as a steeply falling distribution of intensity with minor variations due to the molecular scattering as shown in Fig. 1. These minor variations depend on the molecular parameters, and the basic problem in electron diffraction experiments is to exaggerate

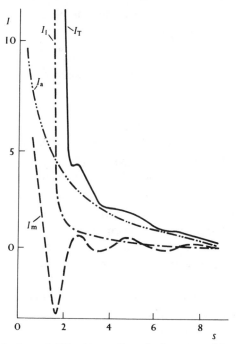

FIG. 1. A total diffraction pattern I_T decomposed into its constituents. (Adapted from L. O. Brockway, *Rev. Mod. Phys.* 1936, **8**, 231.)

these variations in some way so that they become a dominant rather than a subsidiary part of the diffraction pattern.

5.2. The visual method

Although Fig. 1 shows that there are few, if any, real maxima and minima in a diffraction pattern, it is a fortunate physiological accident that the eye reacts to the variable intensity as though the pattern did consist of maxima and minima. In the early days of electron diffraction experiments advantage was taken of this effect to locate the maxima and minima in the diffraction pattern. The relative intensities of the different parts of the pattern were also estimated, and in this way an *experimental scattering curve* could be constructed. In order to interpret this experimental curve, it was customary in visual work to simplify equation (4) in two ways. The variable scattering factor term $F_j F_k/s^4$ was replaced by the product of the atomic numbers $Z_j Z_k$, and the exponential damping term A_{jk} was omitted. The resulting equation for the molecular scattering is now

$$I'_{\mathrm{m}} = k\left\{ \sum_{\substack{j=1 \\ (j \neq k)}}^{N} \sum_{k=1}^{N} Z_j Z_k \frac{\sin(sr_{jk})}{sr_{jk}} \right\}. \qquad (5)$$

The next step in a structure determination is to assume a model for the sample molecule and to select trial values for the internuclear distances r_{jk}. Equation (5) is then used to obtain various *calculated scattering curves* which are compared with the experimental curve. The two simplifications introduced in order to obtain equation (5) will result in slight alterations in the shape of the calculated curves, compared with curves calculated from equation (4), but the positions of the maxima and minima will not be affected. The calculated curve that gives the best fit with the experimental curve is assumed to be the correct one, and the values of r_{jk} which were used to obtain this curve are taken as being characteristic of the sample molecule.

In certain simple cases, of which we may consider silicon tetrafluoride SiF_4, the problem becomes more straightforward and the results more reliable because only one molecular parameter has to be determined. If we assume that SiF_4 is a regular tetrahedron, the molecule has four equal Si—F distances and six equal F—F distances. Hence equation (5) becomes

$$I'_{\mathrm{m}}/k = 6Z_{\mathrm{F}}^2 \frac{\sin\{sr_0(\mathrm{FF})\}}{sr_0(\mathrm{FF})} + 4Z_{\mathrm{Si}} Z_{\mathrm{F}} \frac{\sin\{sr_0(\mathrm{SiF})\}}{sr_0(\mathrm{SiF})}.$$

For a regular tetrahedral molecule $r_0(\mathrm{FF}) = (8/3)^{\frac{1}{2}} r_0(\mathrm{SiF})$ and, since $Z_{\mathrm{F}} = 9$ and $Z_{\mathrm{Si}} = 14$,

$$I'_{\mathrm{m}}/k = 486 \frac{\sin\{(8/3)^{\frac{1}{2}} sr_0(\mathrm{SiF})\}}{(8/3)^{\frac{1}{2}} sr_0(\mathrm{SiF})} + 504 \frac{\sin\{sr_0(\mathrm{SiF})\}}{sr_0(\mathrm{SiF})}. \qquad (6)$$

Equation (6) is now used to calculate values of I'_m for various arbitrary values of $sr_0(SiF)$ between 0 and 50. The results are plotted and the positions of the maxima and minima noted. The calculated curve for SiF_4 is shown in Fig. 2.[†] The electron diffraction photograph is now taken, and from the position of the rings, the value of the specimen to plate distance, and the

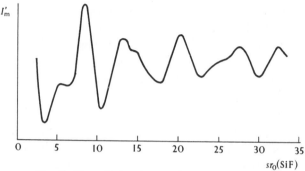

FIG. 2. The calculated scattering curve for SiF_4.

known wavelength of the electrons, it is possible to calculate values of s corresponding to the maxima and minima. At each maximum and minimum, division of $sr_0(SiF)$ by s gives a value for $r_0(SiF)$. The actual results for SiF_4 are shown in Table XIX.

TABLE XIX

Electron diffraction results for SiF_4

Maxima	Minima	$sr_0(SiF)$	s	$r_0(SiF)$
1		8·30	5·406 Å^{-1}	1·535 Å
	2	10·69	6·932	1·542
2		13·09	8·266	1·584
	3	17·82	11·59	1·538
3		20·25	13·02	1·555
	4	22·67	15·08	1·503
4		27·42	17·27	1·588
	5	29·80	19·32	1·542
5		32·20	21·34	1·509
				Mean 1·544 Å

Some of the earlier molecular structures obtained by the visual method have since proved to be unreliable or definitely wrong,

[†] L. O. Brockway and F. T. Wall, *J. Am. chem. Soc.* 1934, **56**, 2373.

with the result that electron diffraction has acquired an undeserved reputation for inaccuracy. As we shall see in the next section, developments in technique and interpretation allow, in certain circumstances, very accurate molecular parameters to be obtained. There are two main reasons for the earlier errors. In the first place scattering due to light atoms, particularly hydrogen, is small, and the effect can be submerged in the scattering due to heavier atoms. In many early determinations hydrogen atoms were ignored entirely, and their omission led to errors in the remainder of the molecular parameters. Secondly, the number of diffraction rings seldom reaches double figures, so that there are less than twenty maxima and minima from which to determine the molecular parameters. For some molecules this proved to be an insufficient margin of safety, and sometimes completely incorrect models were stated to be compatible with the experimental curve. When, however, the shape of the molecule could be assumed with certainty, as in the case of SiF_4, the visual method could give useful values for a limited number of molecular parameters.

5.3. The sector method

In order to eliminate the steeply falling background due to the atomic scattering, it is now usual to rotate a heart-shaped sector above the photographic plate whilst the diffraction pattern is being recorded. By a careful choice of dimensions of the sector most of the steep fall-off can be eliminated and the ripples in the molecular scattering curve then become more pronounced. The intensity of the resulting curve is measured with a microphotometer and, to ensure uniformity, it is advantageous to rotate the photographic plate rapidly about the centre of the diffraction pattern whilst the trace is being recorded. A typical microphotometer trace is shown in Fig. 3. In order to obtain the molecular scattering curve from such a trace, a smooth background curve is drawn through the oscillations on the basis that there must be equal amounts of positive and negative area defined by the molecular scattering curve and the background line.

Nowadays molecular scattering curves are always analysed by a method that leads directly to the internuclear distances. The molecular scattering curve is used to calculate the *radial distribution function*, which is given by

$$f(r) = \int_{s_1}^{s_2} sI_{\mathrm{m}} A' \sin(sr) \, ds,$$

where s_1 and s_2 represent the limits of the scattering angles for which intensity measurements can be made, and A' is an

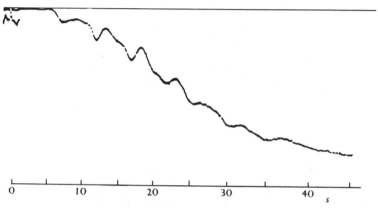

FIG. 3. Microphotometer trace of the diffraction pattern of benzene. (H. Viervoll, *Int. Instrum. Meas. Conf. Stockholm*, 1952, 64.)

exponential damping function of the form $\exp(-as^2)$. The constant a is selected so as to ensure conveniently rapid convergence of the integrand. The evaluation of this integral is quite laborious and is frequently done with automatic computing equipment. The radial distribution function $f(r)$ is related to the probability of finding a particular internuclear distance r in the molecule, so that the calculated function consists of a number of peaks, each of which represents an internuclear distance present in the sample molecule. An example of such a curve is shown in Fig. 4. This curve was calculated from the experimental molecular scattering curve of trifluorochloromethane CF_3Cl.[†] If it is assumed that CF_3Cl has $3m$ (C_{3v}) symmetry, there are

[†] L. S. Bartell and L. O. Brockway, *J. chem. Phys.* 1955, **23**, 1860.

four distinct internuclear distances, and the four peaks corresponding to these four distances are clearly visible in Fig. 4. Values for the bonded distances $r_0(CF)$ and $r_0(CCl)$ can be read off directly from the distribution curve, and the remaining molecular parameter can be calculated from either of the non-bonded distances. The results are given in Table XX together with some other electron diffraction results for halogenomethanes.

TABLE XX

Some electron diffraction results for the molecular parameters of halogenomethanes

Molecule	Bond angle		Bond	r_0
CCl_2F_2	FCF	$109.5 \pm 3°$	CF	$1.33_5 \pm 0.02$ Å
	ClCCl	$108.5 \pm 2°$	CCl	$1.77_5 \pm 0.02$
CH_3Cl	HCH	$110 \pm 2°$	CH	1.11 ± 0.01
			CCl	1.784 ± 0.003
$CClF_3$	FCF	$108.6 \pm 0.4°$	CF	1.328 ± 0.002
			CCl	1.751 ± 0.004
$CBrF_3$	FCF	$109.5 \pm 2°$	CF	1.343 ± 0.021
			CBr	1.911 ± 0.033
CF_3I	FCF	$108.4 \pm 1.9°$	CF	1.340 ± 0.02
			CI	2.135 ± 0.03
CF_4	Tetrahedral		CF	1.323 ± 0.005
CCl_4	Tetrahedral		CCl	1.766 ± 0.005
CBr_4	Tetrahedral		CBr	1.942 ± 0.003

It is rare for the radial distribution curve to show fully resolved peaks as in Fig. 4, and the $f(r)$ curve usually consists of unresolved peaks corresponding to pairs of atoms at approximately the same distance apart. These composite peaks are resolved into atomic peaks on the assumptions that the shape of an atomic peak can be represented by an equation of the form $e^{-x^2/c}$, where c is a constant, i.e. that the curve is Gaussian, and also that the area under an atomic peak formed by two atoms multiplied by the distance between the two atoms is proportional to the product of the scattering powers of the two atoms. Fig. 5 shows the calculated radial distribution curve of benzene decomposed into the various atomic peaks.[†] The three C—C atomic peaks occur at distances of 1·393, 2·410,

† I. L. Karle, *J. chem. Phys.* 1952, **20**, 65.

and 2·786 Å, corresponding to the distances between *ortho*, *meta*, and *para* carbon atoms, and the four C—H peaks occur at 1·08, 2·13, 3·40, and 3·89 Å. These distances are consistent with a regular plane hexagonal molecule with $r_0(CC) = 1·393$ Å and $r_0(CH) = 1·08$ Å. The estimated uncertainty is about $\pm 0·005$ Å in $r_0(CC)$ and $\pm 0·02$ Å in $r_0(CH)$. If these distances are now used to calculate the molecular scattering curve from equation (4), it is possible to vary the damping factors A_{jk} so as to reproduce

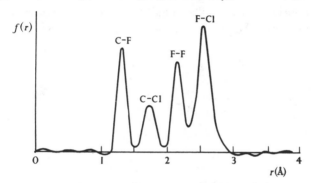

FIG. 4. The radial distribution function of CF_3Cl.

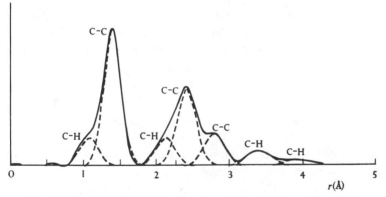

FIG. 5. The radial distribution function of benzene decomposed into its atomic peaks.

very closely the minor variations in the observed scattering curve. From the best values of these damping factors the average amplitudes of vibration of the different pairs at atoms may be obtained. Attempts have been made to use these amplitudes of vibration to correct the apparent internuclear distances to equilibrium distances, but we shall not discuss these corrections since there is as yet no general agreement on the precise significance of r_0 values obtained by electron diffraction.

5.4. The structure of diboron tetrachloride

Another interesting example of the application of electron diffraction is to the determination of the structure of diboron tetrachloride B_2Cl_4. The equilibrium configuration of this molecule might be planar with mmm (D_{2h}) symmetry (Fig. 6 (a)),

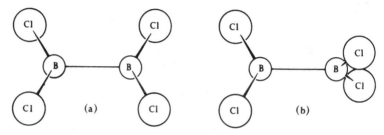

Fig. 6. Two possible shapes for B_2Cl_4: (a) planar, with mmm (D_{2h}) symmetry; (b) BCl_2 groups at right angles, with $\bar{4}2m$ (D_{2d}) symmetry.

or the two BCl_2 groups might be at right angles giving a molecule of $\bar{4}2m$ (D_{2d}) symmetry (Fig. 6 (b)). In either case, since the B—B bond is effectively a single bond, it is probable that the two halves of the molecule may twist through a large angle with respect to each other. In the solid the symmetry is mmm (D_{2h}) as shown by an X-ray diffraction study at $-165°$ C,[†] but a study of the vibration frequencies leads to the conclusion that the correct symmetry in the gas phase is $\bar{4}2m$ (D_{2d}).[‡] This difference between the molecule in the solid and gaseous states has been confirmed by electron diffraction.[§]

The method outlined in the previous section was employed with a nozzle temperature of $-22°$ C. The experimental radial distribution curve is shown at the top of Fig. 7. Below are given

† M. Atoji, P. J. Wheatley, and W. N. Lipscomb, *J. chem. Phys.* 1957, **27**, 106.

‡ D. E. Mann and L. Fano, *J. chem. Phys.* 1957, **26**, 1665.

§ K. Hedberg and R. Ryan, *J. chem. Phys.* 1964, **41**, 2214.

three calculated radial distribution curves for a molecule of $\bar{4}2m$ (D_{2d}) symmetry, a molecule of mmm (D_{2h}) symmetry, and a molecule in which the two halves are rotating freely. Despite

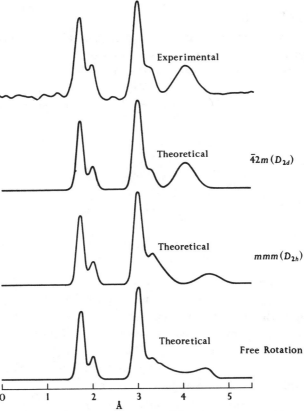

FIG. 7. Radial distribution curves. The theoretical curves include contributions from $SiCl_4$ and BCl_3 impurities estimated at 10 and 30 moles per cent.

the presence of small impurity peaks at 2·02 and 3·30 Å, due to $SiCl_4$ and BCl_3 produced through attack by B_2Cl_4 of the silicone stopcock grease, there can be no doubt that the experimental curve can be reproduced only by a molecule of $\bar{4}2m$ (D_{2d}) symmetry. The fine details of the curve suggest that the two halves of the molecule are oscillating through an angle of about 20° with respect to each other, and that the height of the rotational barrier is about 2·5 kcal/mole.

It is most unusual for a molecule to change from one configuration of high symmetry in the solid phase to another configuration of high, but different, symmetry in the gas.

5.5. Limitations

It was pointed out in section 5.2 that electron diffraction can give accurate values for a limited number of molecular parameters if the symmetry of the molecule is known. However, if

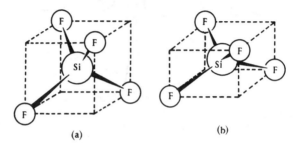

(a)   (b)

FIG. 8. Two conceivable shapes for SiF_4: (a) tetrahedral, with $\bar{4}3m$ (T_d) symmetry, (b) flattened, with $\bar{4}2m$ (D_{2d}) symmetry.

the shape of the molecule is unknown, the results are much less certain. For instance, a regular crown-shaped benzene molecule, with the carbon atoms lying alternately 0·10 Å above and below their mean plane, is equally compatible with the radial distribution curve discussed in section 5.3. The small limits of error quoted are valid only if benzene has $6/mmm$ (D_{6h}) symmetry—which it undoubtedly has. In the same way, if SiF_4 were not a regular tetrahedron (Fig. 8 (a)), but were slightly flattened (Fig. 8 (b)) in such a way as to have $\bar{4}2m$ (D_{2d}) instead of $\bar{4}3m$ (T_d) symmetry, it would not be possible to obtain accurate values for the angles. In general, the higher the symmetry of the molecule the fewer the number of internuclear distances to be determined, and the more accurate the deduced molecular parameters. However, if the symmetry is uncertain, changes in one parameter can be compensated by changes in others, so that a whole series of slightly different molecular structures may be consistent with the molecular scattering curve.

If a molecule possesses more than about six parameters, it is necessary to assume values for some parameters in order to obtain information about the rest of the molecule. This usually reduces the confidence that can be placed in the derived results. Since the molecular scattering curve is least sensitive to the lightest atoms, it is expedient first to assume lengths for bonds

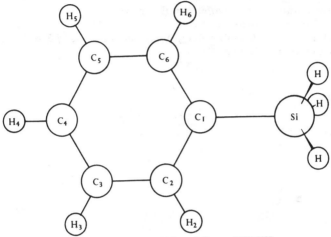

FIG. 9 The molecule of phenylsilane, $C_6H_5SiH_3$.

involving hydrogen atoms and, if necessary, to proceed to the next lightest atom in the molecule. Thus in phenylsilane $C_6H_5SiH_3$ (Fig. 9), which is a very difficult molecule to investigate by electron diffraction because of its size and lack of symmetry, it was found necessary to assume first of all that the angles about the silicon atom were tetrahedral and that $r_0(SiH) = 1.42$ Å.†

TABLE XXI

Molecular parameters of phenylsilane determined by electron diffraction

$r_0(SiC_1)$	$1.84_3 \pm 0.005$ Å
$r_0(C_1C_2)$, etc.	$1.39_2 \pm 0.005$
$r_0(C_2H_2)$, etc.	1.10 ± 0.02
$C_6C_1C_2$	$117.4°$
$C_3C_4C_5$	$120.8°$

† F. A. Keidel and S. H. Bauer, *J. chem. Phys.* 1956, **25**, 1218.

With these given parameters it was possible to deduce the results shown in Table XXI. It will be seen that by no means all of the many molecular parameters were determined, but the gross features of the molecule, including some angular distortion in the phenyl ring, were established.

REFERENCES

R. BEECHING, *Electron Diffraction* (Methuen).

Z. G. PINSKER, *Electron Diffraction* (Butterworth).

M. H. PIRENNE, *The Diffraction of X-rays and Electrons by Free Molecules* (Cambridge University Press).

VI

X-RAY DIFFRACTION: GENERAL CONSIDERATIONS

6.1. Introduction

ALTHOUGH X-ray diffraction can be used to study gases, liquids, and solutions, by far the most important application is to the investigation of the solid state, to which we shall restrict this discussion. In recent years somewhere between five and ten times as many molecular structures have been determined by means of X-ray diffraction as by all other experimental methods put together. This is understandable when we remember that all substances can be obtained in a solid form whereas by no means all can be obtained as gases. We shall see later that the molecular parameters obtained by X-ray diffraction are, except in a small number of special cases, less accurate than those obtained by most other methods. Nevertheless, because of its wide applicability to both small and large molecules, X-ray diffraction is of paramount importance for the determination of molecular symmetry and molecular parameters.

Since we are now interested in solids, we must, before we can understand X-ray diffraction, become familiar with a whole range of new concepts that do not apply when molecules are isolated in the gas phase. The present chapter will be concerned largely with the introduction of these ideas. In the following two chapters we shall see how X-ray diffraction can be applied first to the determination of molecular symmetry and then to the determination of molecular parameters.

6.2. The generation of X-rays

X-rays are produced when electrons with sufficient kinetic energy strike the atoms of any element. An X-ray tube consists of a filament or cathode and a metal target or anode. The filament, usually made of tungsten wire, is heated to produce a beam of electrons which is accelerated by a high voltage (~ 50 kV) on to the target. The X-ray tube must be evacuated to prevent absorption of the electrons by gas molecules, and the target must be water-cooled, otherwise the electron beam would burn a hole in it. The X-rays pass out from the tube through windows which must be strong enough to hold a vacuum, but which must absorb X-rays only slightly. Since the absorption factor of atoms for X-rays increases with the atomic number, the windows are usually made from thin sheets of beryllium or aluminium.

The resulting X-rays have a broad spectrum of wavelengths, but superimposed on this *white radiation* are a few sharp peaks of *characteristic radiation*. The latter, as the name implies, are characteristic of the metal from which the target is made. It is possible to filter out all but one main peak called the K_α radiation, and X-ray diffraction experiments are almost always carried out with this highly monochromatic K_α radiation. By an appropriate choice of target material the wavelength of the X-rays may be varied, but the most commonly used metals are copper and molybdenum for which the wavelengths of the K_α radiation are 1·542 and 0·711 Å respectively.

6.3. Crystals

The main feature that distinguishes most crystals from other condensed states of matter such as liquids or amorphous solids is the anisotropy of the physical properties. Although many crystals do have well-defined shapes and sharp edges, these are not good criteria of crystallinity because they can be obscured by grinding or polishing and yet the material is still crystalline. The shape of a crystal, or the crystal *habit* as it is often called, depends on the environmental factors controlling the growth of the crystal and on its subsequent treatment. Thus crystals of the same substance grown under different conditions can present very different appearances. Nevertheless, it was observed long ago that, no matter what the crystal habit, the angles between corresponding faces of crystals of the same substance are constant. Practically all solids are crystalline. We know now that the crystallinity is due to the fact that the discrete atoms, ions, or molecules of which a crystal is usually composed normally try to pack as close together as possible. In doing this they find that the system of lowest potential energy is one in which the units are arranged periodically in three dimensions with the environment of each unit identical. It is this periodic arrangement of units that accounts for the constancy of the interfacial angles.

In fact not many crystals are quite as regular as the above paragraph implies. A macroscopic crystal usually consists of a

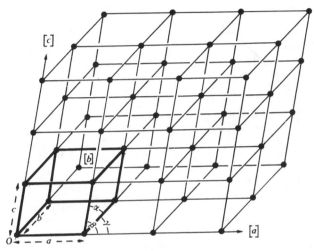

FIG. 1. A space lattice of points.

large number of crystallites each of which is a true single crystal, but the crystallites are displaced with respect to each other. The displacement is normally quite slight, and for many purposes we can ignore it, so that a macroscopic crystal often has large faces and always a characteristic set of interfacial angles, even though it may not be a true single crystal.

In order to discuss a crystal, it is convenient to forget for the moment that it is made up of atoms or molecules, and to replace each unit by a point. By repeating these points in the same way that the atoms or molecules are repeated in the crystal, we build up a *space lattice* of points (Fig. 1). The lattice (and therefore the crystal) can be defined in terms of three non-parallel axes x, y, and z, and of the angles between these axes. Along each axis a point will repeat at a distance that we call a, b, and c respectively.

a, *b*, and *c* are known as the *unit translations*. The angle between *x* and *y* is called γ, that between *y* and *z* is called α, and that between *x* and *z* is called β. It is customary, when referring to crystallographic axes, to use only the letters *a*, *b*, and *c*, and, in order to avoid confusion with the translations, to enclose the letters in square brackets. Thus [a] is to be read as 'the *a*-axis',

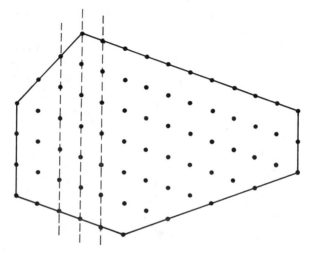

Fig. 2. Relation between crystal faces and density of lattice points.

that is the *x*-axis along which the unit of translation is *a*. Although there are numerous ways in which we can choose our axes, that is to say, there are numerous sets of axes which define the same space lattice, we usually try to choose *a*, *b*, and *c* to be as small as possible, and α, β, γ to be as near 90° as we can.

The planes that occur as crystal faces are those which contain a high density of lattice points (Fig. 2). The method we use to define these planes is based on the intercepts that the planes make on [a], [b], and [c]. If for simplicity we consider a two-dimensional lattice (Fig. 3) and connect rows of points by the lines *AA*, *BB*, and *CC*, then these lines represent possible faces of the crystal. In order to describe these faces, we read off the intercepts on [a], [b], and [c] and take the reciprocals of the multiples of *a*, *b*, and *c*. Since we are dealing with a two-dimensional lattice, the intercept on the third axis ($[b]$ in Fig. 3) will be infinite and the reciprocal will be zero. We next clear of fractions by multiplying by the lowest common multiple (excluding, of course, infinity) and we obtain three numbers known as the *Miller indices* of the plane. These operations are shown in detail in Table XXII. Exactly the same procedure is

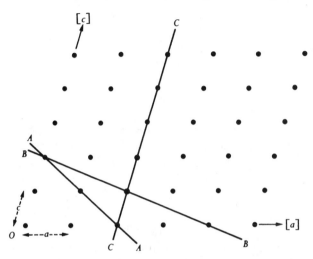

FIG. 3. The representation of crystal faces by Miller indices.

TABLE XXII

The generation of Miller indices

Face	Intercepts	Reciprocals of multiples	Clear of fractions	Miller indices (hkl)
AA	$2a, \infty b, 2c$	$1/2, 1/\infty, 1/2$	$\times 2$	101
BB	$4a, \infty b, 2c$	$1/4, 1/\infty, 1/2$	$\times 4$	102
CC	$2a, \infty b, \infty c$	$1/2, 1/\infty, 1/\infty$	$\times 2$	100

followed in three dimensions. Each Miller index triplet represents a series of parallel equi-spaced planes containing all the lattice points as shown by the dotted lines in Fig. 2. It will be seen from Table XXII that planes parallel to one of $[a]$, $[b]$, or $[c]$ have one zero index, whilst those that are parallel to two of these axes have two zero indices.

In crystallography we often wish to discuss the perpendicular distance d_{hkl} between adjacent members of the set of parallel planes represented by the Miller indices (hkl). This distance can be worked out by quite straightforward geometry, but the answer is a complicated expression if the axes are not orthogonal. If $\alpha = \beta = \gamma = 90°$, then

$$\frac{1}{d_{hkl}^2} = \frac{h^2}{a^2} + \frac{k^2}{b^2} + \frac{l^2}{c^2} \tag{1}$$

which, when $a = b = c$, reduces to

$$d_{hkl} = \frac{a}{\sqrt{(h^2 + k^2 + l^2)}}. \tag{2}$$

6.4. The crystal systems

It can be seen from Fig. 1 that a crystal may be regarded as being composed of a large number of blocks of identical size and shape. One block is known as a *unit cell*. In order to satisfy our criterion that the environment of every lattice point must be the same, it follows that the blocks must be capable of repetition in space without leaving any gaps. We have already seen in Chapter I that this criterion sets a limit on the possible symmetry about a point, and it also sets a limit on the number of types of unit cell that can be used to build a crystal. The thirty-two crystallographic point groups listed in Table I give rise to only seven different shapes of unit cell, and these are known as the seven *crystal systems*. In Table XXIII we have rearranged the point groups in terms of the crystal systems. We have also shown the minimum symmetry required for each particular system and the shape of each type of unit cell, the shape being a consequence of these symmetry requirements. In this table the symbol $\neq$ is to be interpreted as meaning 'not necessarily equal to'. By accident two quantities may be equal, but if this accidental equality occurs the arrangement of matter within the crystal will still be of the lower symmetry, and it is the symmetry that really counts. For instance, it may happen that all three unit cell dimensions of a tetragonal crystal are equal within experimental error. We would then say that the crystal is *pseudo-cubic*, but there would be many properties of the crystal that would demonstrate the

TABLE XXIII

The seven crystal systems

Point group symbol		System	Unit cell	Minimum symmetry requirements
S	H–M			
C_1	1	Triclinic	$\alpha \neq \beta \neq \gamma \neq 90°$	None
S_2	$\bar{1}$		$a \neq b \neq c$	
C_2	2	Monoclinic	$\alpha = \gamma = 90°$	One twofold axis or one
C_h	m		$\beta \neq 90°$	mirror plane
C_{2h}	$2/m$		$a \neq b \neq c$	
D_2	222	Orthorhombic	$\alpha = \beta = \gamma = 90°$	Any combination of three mutually perpendicular twofold axes or mirror planes
D_{2h}	mmm		$a \neq b \neq c$	
C_{2v}	mm			
C_3	3	Trigonal	(Rhombohedral axes) $\alpha = \beta = \gamma \neq 90°$ $a = b = c$ (Hexagonal axes) $\alpha = \beta = 90°,$ $\gamma = 120°$ $a = b \neq c$	One threefold axis
D_3	32			
S_6	$\bar{3}$			
D_{3d}	$\bar{3}m$			
C_{3v}	$3m$			
C_6	6	Hexagonal	$\alpha = \beta = 90°$ $\gamma = 120°$ $a = b \neq c$	One sixfold axis or one sixfold inversion axis
C_{3h}	$3/m$			
C_{6h}	$6/m$			
D_6	62			
D_{3h}	$\bar{6}2m$			
D_{6h}	$6/mmm$			
C_{6v}	$6mm$			
C_4	4	Tetragonal	$\alpha = \beta = \gamma = 90°$ $a = b \neq c$	One fourfold axis or one fourfold inversion axis
C_{4h}	$4/m$			
D_4	42			
D_{4h}	$4/mmm$			
S_4	$\bar{4}$			
D_{2d}	$\bar{4}2m$			
C_{4v}	$4mm$			
T	23	Cubic	$\alpha = \beta = \gamma = 90°$ $a = b = c$	Four threefold axes at $109° \, 28'$ to each other
O	43			
T_h	$m3$			
O_h	$m3m$			
T_d	$\bar{4}3m$			

absence of the threefold axes which are necessary for a truly cubic crystal. The seven different types of unit cell are shown in Fig. 4.

One of the first tasks that a crystallographer has to do is to find the size and shape of the unit cell. Although it is often possible, by measurement of the interfacial angles, to determine

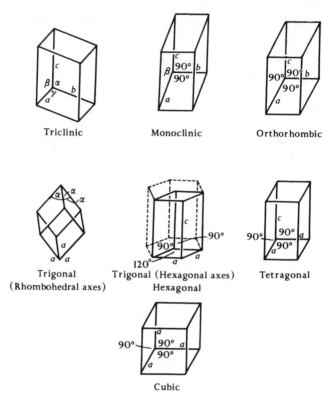

Triclinic   Monoclinic   Orthorhombic

Trigonal   Trigonal (Hexagonal axes)   Tetragonal
(Rhombohedral axes)   Hexagonal

Cubic

FIG. 4. The seven crystal systems.

the external symmetry of the crystal, and hence to assign it to the proper crystal system, it is necessary to use X-rays in order to determine the absolute lengths of the sides of the unit cell. We shall see in section 6.6 how this may be done.

6.5. Bravais lattices

A crystal is built up by the repetition by translation of the contents of the unit cell. In Fig. 5 we have drawn an array of our coins to represent a crystal. For simplicity we still work in two dimensions. We have outlined the simplest cell by full lines, the unit cell translations being a and c. The unit cell contains one complete repeat unit. Such a cell is known as a *primitive cell*. Once we have selected the axes to define our unit cell, it is fixed in shape, size, and orientation. However, it is not fixed in position, or, in other words, the choice of origin is arbitrary,

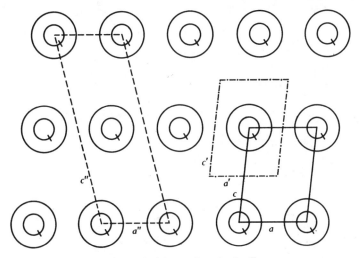

FIG. 5. Primitive and centred cells.

and we could equally well have defined our unit cell by the lines a' and c'. In crystallography an origin is often selected at one particular point rather than at any other, but this is just a matter of convenience and not of necessity. As stated in section 6.3, we could have selected other translations to define our unit cell, but we have adhered to the rule that they should be the shortest possible and that the angle between them should be close to 90°. Any other primitive cell will have the same volume (area, in two dimensions) as the one we have chosen, even though the axes may be of very different lengths.

However, it is not absolutely necessary to choose a primitive cell. We could choose the cell defined by the lines a'' and c'' in Fig. 5, but we now have a *non-primitive* or *centred cell* containing two repeat units and of twice the volume. Indeed it is often preferable to choose a centred cell. In Fig. 6 another array of coins is drawn and a centred cell defined by the lines a and c. This is an orthogonal cell, unlike the primitive cell outlined by the lines a' and c'. In other words, the non-primitive cell displays the full symmetry of this array, whereas the primitive cell masks it. We could, of course, choose a non-primitive cell

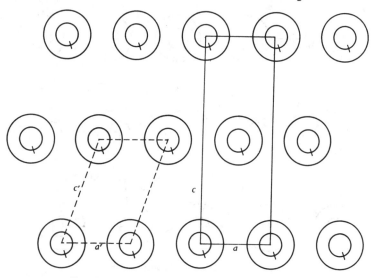

FIG. 6. Retention of symmetry by a centred cell.

with any number of repeat units, but this is unnecessary and, in three dimensions, it is found that only three types of non-primitive lattice are required to preserve the symmetry of an array. These are the *body-centred lattice*, denoted by the symbol I, in which there is an additional lattice point at the centre of the unit cell; the *side-centred lattice* with one pair of opposite faces centred, denoted by A, B, or C depending on which pair of faces is centred; and the *face-centred* lattice with all faces centred, denoted by F. These centred lattices are summarized in Table XXIV, together with the number of repeat units in each type of lattice.

It is not possible for each of the seven types of unit cell to have each of the three types of centring. For instance, a cubic cell cannot have only one pair of faces centred because then the

TABLE XXIV

Primitive and centred lattices

Name	Symbol	No. of repeat units in cell
Primitive	*P*	1
Body-centred	*I*	2
Side-centred	*A* or *B* or *C*	2
Face-centred	*F*	4

symmetry would no longer be cubic. A triclinic cell cannot have any type of centring, because it would always be possible to choose a smaller primitive cell which would still, of course, be triclinic, since this is the lowest possible symmetry. If we include the different possible sorts of centring, we find that there are now fourteen lattices distributed amongst the seven crystal systems, and these are known as the fourteen *Bravais lattices*. They are shown in Fig. 7. It is always possible to choose a primitive cell from a centred lattice, but it is usually much more convenient to select a unit cell which shows the full symmetry of the array.

After a crystallographer has determined the shape and size of the unit cell, his next task is to discover whether the lattice is primitive or centred. We shall see how this may be done in section 6.10.

6.6. Diffraction of X-rays and the determination of the cell constants of single crystals

If a parallel beam of monochromatic light, represented by AB and DE in Fig. 8, is directed perpendicularly on to a row of equispaced points, most of the light will pass straight on but some will be diffracted if the distance between the points is of the same order of magnitude as the wavelength of the light. The diffracted rays will interfere unless the difference between the path-lengths is zero or an integral number of whole wavelengths. In Fig. 8 the ray EF is a distance $a\cos\phi$ behind the ray BC, and the rays will reinforce only when

$$n\lambda = a\cos\phi. \tag{3}$$

Since ϕ is a constant for particular values of n and λ, the diffracted

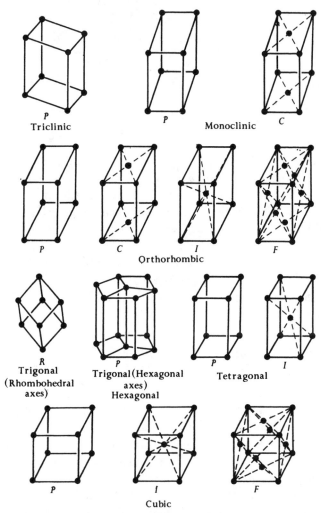

FIG. 7. The fourteen Bravais lattices.

rays will form two cones, one on each side of the incident beam. When $n = 0$, the two zero-order cones combine to form a plane surface as shown in Fig. 9. When $n = \pm 1$, the first-order diffracted rays form two cones with the row of points as axis. As n increases the cones get successively narrower. This simple picture of diffraction by a one-dimensional array of points can be generalized to three dimensions and for oblique incidence of the light, but for our present purposes the simple treatment is adequate.

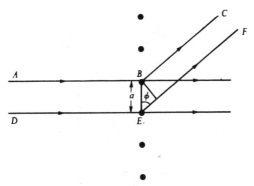

FIG. 8. Diffraction of light by a row of equi-spaced points.

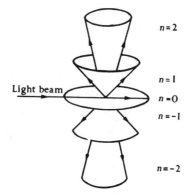

FIG. 9. Cones of rays diffracted by a row of equi-spaced points. (Adapted from C. W. Bunn, *Chemical Crystallography*, Clarendon Press.)

Suppose we mount a crystal with one of the principal axes, say $[a]$, vertical and we shine a beam of X-rays on to it. Along $[a]$ there are identical diffracting units with a spacing a. This spacing and the wavelength of the X-rays are both of the order of 1 Å, so that diffraction will occur and a series of cones will be formed as described above. If we surround the crystal with a cylinder of film of radius R, a series of lines known as *layer lines* will be formed where the cones cut the film (Fig. 10). In fact, the layers are not continuous but are broken up into discrete spots as we shall see in the next section, but that need not worry

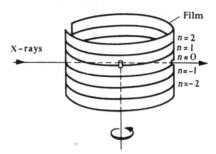

Fig. 10. The formation of layer lines. (Adapted from J. M. Bijvoet, N. H. Kolkmeijer, and C. H. MacGillavry, *X-ray Analysis and Crystals*, Butterworth.)

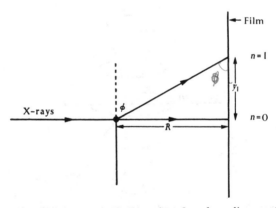

Fig. 11. Determination of cell dimensions from layer line spacings.

us for the moment. If the distance between the zero layer and the first layer is y_1, then (Fig. 11)

$$\tan\phi = R/y_1.$$

Thus we can obtain a value for ϕ and, since $n = 1$ and we know the wavelength of the X-rays, we can substitute in equation (3) and obtain a value for a, the unit of translation in the vertical direction. We can check our value of a by measuring $y_2, y_3, \ldots,$

the distance from the zero layer to the second, third,... layer, and taking $n = 2, 3,...$ in equation (3). We can then alter the orientation of the crystal and set other axes vertical until we have determined sufficient lengths to define completely the size and shape of the unit cell.

6.7. The Bragg equation

An alternative way of regarding diffraction by a set of crystal planes has been proposed by Bragg. In many ways the Bragg

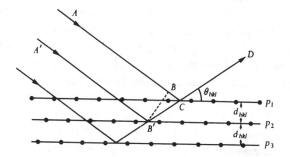

FIG. 12. Reflexion of light from a series of equi-spaced planes.

approach is more illuminating and we shall refer to it frequently in the following sections. Suppose we shine a parallel mono-chromatic beam of X-rays, represented by AB, $A'B'$ in Fig. 12, on to a crystal consisting of a series of planes $p_1, p_2, p_3,...$ at a distance d_{hkl} apart. Some of the X-rays may be regarded as being reflected from the point C on the plane p_1. Other rays will be reflected by successive layers $p_2, p_3,...$. For the reflected beams to emerge as a single beam of appreciable intensity they must reinforce each other. For the reflected beam $B'D$ to reinforce CD, the wave represented by $A'B'D$ must arrive in phase with the wave ACD, and the path difference must be a whole number of wavelengths. Hence

$$B'C - BC = n\lambda.$$

But $B'C = d_{hkl}/\sin\theta_{hkl}$ and

$$BC = B'C\cos 2\theta_{hkl} = d_{hkl}\cos 2\theta_{hkl}/\sin\theta_{hkl}.$$

Therefore

$$\frac{d_{hkl}(1-\cos 2\theta_{hkl})}{\sin \theta_{hkl}} = \frac{d_{hkl}\, 2 \sin^2\theta_{hkl}}{\sin \theta_{hkl}} = n\lambda,$$

and $$n\lambda = 2d_{hkl} \sin \theta_{hkl}.$$ (4)

This is known as the *Bragg equation* and from it we conclude that, for a given crystal with particular spacings d_{hkl}, reflexion occurs only at particular values of the *Bragg angle* θ_{hkl}. In other words, unlike ordinary visible light reflected from a polished surface, the incident beam will not be reflected at any angle of incidence but only when some set of planes is arranged at the appropriate angle to the primary beam. This is the reason why the layer lines discussed in the previous section are not continuous but are broken into discrete spots. Each spot may be regarded as arising as a result of reflexion from some given set of planes, and in order to increase the number of sets that will at some time be in the reflecting position, it is customary to rotate the single crystal round the vertical axis as shown in Fig. 10. For this reason the resulting photograph is known as a *rotation photograph*. A typical example is shown in Fig. 13.

It should be noticed that we are not really entitled to speak of the diffracted beams as reflexions, but the Bragg approach is so useful that this terminology is almost always employed and we shall continue to use it.

We may also conclude from the Bragg equation that there must be a definite relation between the shape and size of the unit cell and the angles at which detectable reflexion occurs, since, for a given wavelength, these angles depend only upon the various interplanar spacings d_{hkl}. If we can measure the Bragg angles, we can calculate the corresponding values of d_{hkl}. All we have to do then in order to obtain a complete picture of the size and shape of the unit cell is to decide from which set of crystal planes each reflexion arises. This involves the allocation of the correct indices (hkl), and the operation is known as *indexing the reflexions*. In the next section we shall discuss one of the various ways in which this indexing may be done.

According to the Bragg equation there are different orders of

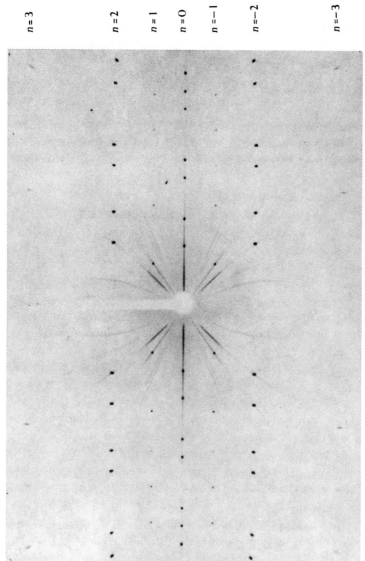

$n = 3$

$n = 2$

$n = 1$

$n = 0$

$n = -1$

$n = -2$

$n = -3$

Fig. 13. A rotation photograph of sodium chloride.

reflexion corresponding to the values $n = 0, 1, 2,...$. The third-order reflexion from the (100) planes can be looked at in this way, or it may be regarded as a first-order reflexion from a set of planes (300) with one-third the spacing of the (100) planes. Consequently we can write the equation for the third-order reflexion either as

$$3\lambda = 2d_{100} \sin \theta_{300},$$

or as

$$\lambda = 2d_{300} \sin \theta_{300}.$$

It is more convenient to choose the second of these alternatives and from now on we shall omit the n from the Bragg equation, knowing that we have allowed for it by choosing the correct spacing.

6.8. The powder method

In Fig. 14 AS represents a collimated beam of X-rays and a powdered crystal is placed at S. As for a single crystal (section 6.6) most of the beam passes straight through the sample in the direction SB, but some of the X-rays will be diffracted. In the fine powder a large number of small single crystals are oriented in all possible directions so that, at any one time, there will be crystals in the correct orientation for diffraction to occur from every possible set of crystal planes. Let us confine ourselves to the (100) and (110) planes. The crystals which are in the correct orientation for reflexion from the (100) planes to occur will produce a circular cone of rays with a half angle $2\theta_{100}$, where θ_{100} is the Bragg angle for the (100) planes. Similarly, a different set of crystals will be in the correct orientation for reflexion from the (110) planes to occur, and these planes will produce a cone of rays with half angle $2\theta_{110}$, where θ_{110} is the Bragg angle for the (110) planes. If the sample is enclosed by a strip of film bent into a circle of radius R, a pair of arcs CC, DD will be formed where each cone cuts the film. A hole is punched in the film at B, otherwise the strong primary beam would cause complete blackening of the film. If we now measure the distance of each arc from the centre of the film, the value of the Bragg angle in degrees is given by

$$\frac{2\theta_{100}}{360} = \frac{y_1}{2\pi R} \quad \text{and} \quad \frac{2\theta_{110}}{360} = \frac{y_2}{2\pi R}. \quad (5)$$

Knowing the Bragg angle and the wavelength of the X-rays, we can calculate the interplanar spacing from equation (4).

We still have to answer the question: How do we know which reflexion came from the (100) and which from the (110) planes? If there were only these two reflexions present, it is clear that the one nearer the centre of the film would have to come from

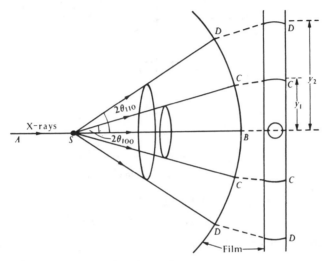

Fig. 14. The formation of a powder photograph.

the (100) planes, since d_{100} must always be greater than d_{110} and, from the Bragg equation, θ_{100} must be less than θ_{110}. But in general there are many reflexions present and we must have a system for assigning the correct indices to each reflexion.

Suppose we have a primitive cubic crystal structure consisting of identical atoms placed at the corners of a cube. In Fig. 15 the (100) planes are defined by the atoms $XXYY$ or equivalent sets, and the (110) planes by $XXZZ$. For a cubic lattice we may combine equation (2) for the interplanar spacing with the Bragg equation to give

$$\sin^2\theta_{hkl} = \frac{\lambda^2}{4a^2}(h^2+k^2+l^2). \qquad (6)$$

Now $\lambda^2/4a^2$ is a constant and $(h^2+k^2+l^2)$ must necessarily be an integer. Furthermore, not all integers can be expressed in the form $(h^2+k^2+l^2)$. In Table XXV the first few allowed values are shown. If we replot the observed X-ray pattern of a simple cubic crystal as intensity of diffraction against $\sin^2\theta_{hkl}$, we shall find six equi-spaced lines with the seventh missing, since $(h^2+k^2+l^2)$ cannot equal seven. Similarly, the 15th, 23rd, 28th, 31st, 39th,... lines will be missing. Consequently not only can

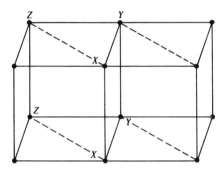

FIG. 15. A simple cubic crystal.

TABLE XXV
Possible values of $(h^2+k^2+l^2)$

(hkl)	100	110	111	200	210	211	220	$\left.\begin{matrix}221\\300\end{matrix}\right\}$	310	311	222
$(h^2+k^2+l^2)$	1	2	3	4	5	6	8	9	10	11	12

we readily identify a primitive cubic crystal, but by inspection we can assign indices (hkl) to each line of the powder photograph. Once we have assigned the indices and have measured each θ_{hkl}, we can calculate the length of the cell side from equation (6).

Powder photographs can be indexed without much difficulty provided the symmetry of the unit cell is sufficiently high, that is provided there are only two cell constants to be determined. This means that we can deal with cubic, tetragonal, hexagonal, and trigonal crystals without too much trouble. However, for crystals of lower symmetry a powder photograph consists of a large number of lines which often overlap and indexing proves to be difficult if not impossible. It is much more convenient, therefore, to use single crystals rather than powders, and we must now see how it is possible to index single crystal photographs.

6.9. The indexing of single crystal photographs

A rotation photograph has three advantages over a powder photograph. In the first place, the reflexions are disposed along a number of lines instead of just one. Consequently the density of spots along any one layer line must be less than the density of lines in the corresponding powder photograph, and the possibility of overlap will be diminished. Secondly, we already know one index of each reflexion. Thus if $[a]$ is vertical, all reflexions on the zero layer have $h = 0$, those on the first layer have $h = 1$, and so on. Thirdly, as discussed in section 6.6, we can determine the size of the unit cell without knowing the indices of the reflexions, no matter what the crystal system. This third point gives a clue to the method used for the indexing of rotation photographs, a method that is the reverse of that described in the last section. From the known cell dimensions we calculate all possible values of d_{hkl} and then θ_{hkl}. From the values of θ_{hkl} and the radius of the camera we can calculate the distance from the centre of the film to each reflexion (hkl). We know from the indices on which layer line each reflexion must lie, so that all we have to do is to find a reflexion on the correct layer line corresponding to each of these distances. Various graphical methods have been devised for the facilitation of this process, and for small unit cells there is little difficulty in indexing all the reflexions.

However, in the same way that the powder method loses its power when the symmetry gets too low, the rotation method loses its power when the unit cell becomes too large. The reflexions on each layer line become so close together that they often overlap and it becomes impossible to sort them out. This difficulty can be partially overcome by oscillation of the crystal through a narrow angular range such as 5° or 15° instead of rotation through 360°. When this is done the layer lines consist of fewer spots since fewer sets of planes have been turned into the reflecting position. A 15° oscillation photograph is compared with the corresponding rotation photograph in Fig. 16. Even though the oscillation range is made very small there is still the possibility that some reflexions will overlap, and it would clearly

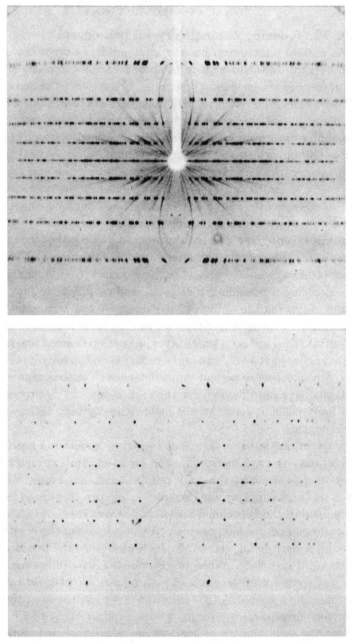

FIG. 16. Rotation and 15° oscillation photographs of sucrose.

be advantageous if some method could be devised which would prevent this.

The most common method of separating different reflexions is by some sort of moving-film technique such as the one devised by Weissenberg. In this method one particular layer line is

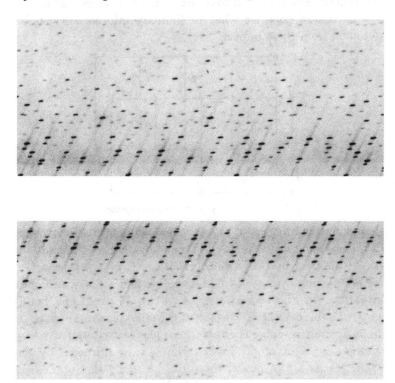

FIG. 17. Weissenberg photograph of $C_{21}H_{21}N_9S_2Br$.

isolated by means of metal screens. Then, as the crystal is rotated, the film is translated parallel to the axis about which the crystal is rotated. Since no two sets of planes can be in the reflecting position at the same time, two spots cannot overlap because the film will have moved between one reflexion and another. The rotary and translatory motions are coupled together so that the spots in any one layer line can be spread right across the film as shown in Fig. 17. It can be seen that the spots lie on regular lines or curves, and it is possible to index a Weissenberg photograph by inspection.

6.10. Determination of the Bravais lattice

We have seen how it is possible to take X-ray photographs of a crystalline substance, and to index all the reflexions that occur. We must now consider what preliminary information can be obtained from these indices.

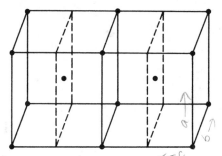

FIG. 18. A body-centred cubic crystal.

A primitive cubic crystal gives reflexions from all possible sets of crystal planes, that is, for all possible values of h, k, and l up to the limit set by the wavelength of the X-rays. If, however, we are dealing with a centred lattice, we find a rather different situation. In Fig. 18 we have drawn two unit cells of a body-centred cubic crystal. If we look at the (100) planes, we see that, unlike primitive cubic crystals, these planes are interleaved by another layer of atoms. When the X-rays diffracted by the (100) planes are in phase, those diffracted by the interleaved atoms will be of exactly opposite phase, since the interleaved layers are half-way between each pair of (100) planes. A crystal consists of very many unit cells and, when the number of cells is large, there are as many atoms in the interleaved layers as there are in the (100) planes. Consequently, the (100) reflexion will be entirely absent. The (200) reflexion, however, will be present because all the atoms lie in the (200) planes, which have, of course, half the spacing of the (100) planes. Similarly, the (110) reflexion will be present because all the atoms lie in the (110) planes. Working in this way we can show that, whenever the sum of the three indices h, k, and l is odd, there is a layer of interleaved atoms, and the reflexion will be absent. These systematic absences are known as *general absences*.

For a face-centred lattice (Fig. 19) we can see that reflexions both from the (100) and (110) planes will be absent and, in

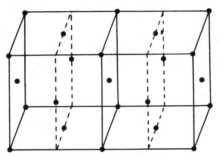

FIG. 19. A face-centred cubic crystal.

general, we find that systematic absences will occur whenever $(h+k)$ or $(k+l)$ or $(h+l)$ is odd. This means that reflexions will occur only if all three indices are odd or if all three are even. In other words, reflexions with mixed indices will be absent.

Although we have considered only cubic lattices, centring leads to the same general absences amongst the (hkl) reflexions no matter what the crystal system. These general absences are summarized in Table XXVI. Once we have indexed all the

TABLE XXVI

Systematic absences due to centring

Type of lattice	Systematic absences
Primitive (P)	None
Body-centred (I)	Reflexions with $(h+k+l)$ odd
A side-centred (A)	Reflexions with $(k+l)$ odd
B side-centred (B)	Reflexions with $(h+l)$ odd
C side-centred (C)	Reflexions with $(h+k)$ odd
Face-centred (F)	Reflexions with mixed indices, i.e. h, k, and l must all be odd or all even for reflexions to be present

reflexions from a crystal we can draw up a list and determine the Bravais lattice by finding what reflexions, if any, are absent from the list.

6.11. Determination of the crystal structure of sodium chloride

After this preliminary survey, we are now in a position to consider the structural determination of a simple ionic crystal such as sodium chloride. Suppose we use copper radiation with a wavelength of 1·542 Å to take an X-ray powder diagram of NaCl in a cylindrical camera of known radius. A typical result is shown in Fig. 20. After developing and drying the photograph we measure the distance between each pair of corresponding arcs on the film, and halve this reading to give the distance of each arc from the centre of the film. We can then calculate the various Bragg angles from equation (5). From the various values of θ_{hkl} we can obtain values of $\sin^2\theta_{hkl}$, and inspection shows that they are, within experimental error, divisible by the same number. The mean value of this highest common factor is 0·01875. The presence of this common factor shows that the unit cell must be defined in terms of only one parameter and is therefore cubic. From equation (6) this factor must be equal to $\lambda^2/4a^2$, which yields a value of $a = 5·631$ Å. We next divide each value of $\sin^2\theta_{hkl}$ by 0·01875 and obtain a series of integers which must be equal to $(h^2+k^2+l^2)$. We may then select the appropriate values of h, k, and l. These various operations are shown in Table XXVII. From these values of the indices

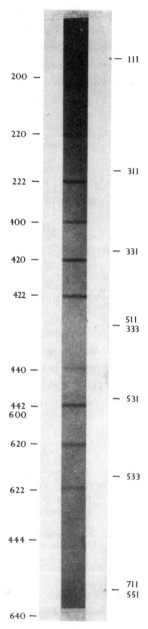

Fig. 20. A powder photograph of sodium chloride.

of the observed reflexions, which have been inserted in Fig. 20, we see that only those reflexions are present which have all three indices odd or all three even. We may conclude, therefore, that NaCl is face-centred cubic.

TABLE XXVII

X-ray powder results for sodium chloride†

θ_{hkl}	$\sin^2\theta_{hkl}$	$(h^2+k^2+l^2)$	(hkl)
13° 41′	0·0560	3	111
15° 51′	0·0746	4	200
22° 44′	0·1492	8	220
26° 56′	0·2052	11	311
28° 14′	0·2239	12	222
33° 07′	0·2984	16	400
36° 32′	0·3544	19	331
37° 39′	0·3731	20	420
42° 00′	0·4477	24	422
45° 13′	0·5036	27	511, 333
50° 36′	0·5972	32	440
53° 54′	0·6529	35	531
55° 02′	0·6715	36	600, 442
59° 45′	0·7462	40	620
63° 35′	0·8022	43	533
64° 57′	0·8208	44	622
71° 08′	0·8954	48	444
77° 15′	0·9513	51	711, 551
80° 02′	0·9700	52	640

† Adapted from *Internationale Tabellen zur Bestimmung von Kristallstrukturen* (Borntraeger).

Next we must determine how many formula units of NaCl there are in the cell. In order to do this we need to know the density of the crystals. The simplest method of finding the density of an ionic crystal is to make up mixtures of non-polar liquids until one is prepared in which the crystals neither rise nor fall. The density of the liquid mixture must then be identical with that of the crystals and may be determined pyknometrically or by calculation from the known densities of the original liquids. The value found in this rather crude way for NaCl is 2·16 g/cm³. Since the formula weight of NaCl is 58·61 (on the physical scale), the molar volume is therefore $58·61/2·16 = 27·13$ cm³. The volume per 'molecule' is then

$$\frac{27·13}{6·025 \times 10^{23}} \text{ cm}^3 = 45·0 \text{ Å}^3.$$

The volume of the unit cell is 178·5 Å³, so that there must be four formula units of NaCl in the cell. One possible way in which these four units can be placed in a face-centred cubic cell is shown in Fig. 21. We shall discuss in section 8.3 how this structure may be confirmed.

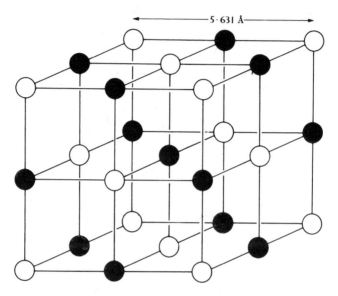

FIG. 21. The crystal structure of sodium chloride.

It will be seen that there are no molecules in this crystal, but that each sodium ion is surrounded by six chloride ions, and vice versa. In Fig. 21, therefore, the full circles can represent either sodium or chloride ions. It may appear from this figure that there are more than four units of NaCl in this cell, but we must remember that a circle at the corner of the cell is shared between eight cells, one on an edge between four cells, and one at the centre of a face between two cells. The only circle that belongs solely to the cell is the one right at the centre. If we consider first the full circles in Fig. 21, we see that there is a total of $1 + 12 \times \frac{1}{4} = 4$ in the cell. There are $8 \times \frac{1}{8} + 6 \times \frac{1}{2} = 4$ empty circles. This gives the required four formula units.

VII

X-RAY DIFFRACTION: DETERMINATION OF MOLECULAR SYMMETRY

7.1. Symmetry elements involving translation

W E saw in section 6.10 how lattice centring leads to systematic absences and how we can determine the Bravais lattice from these absences. It is often found, however, that even though the lattice is primitive there are still some systematic absences. These *special absences* can also give information about the distribution of matter within the unit cell, and we must now consider how they arise.

It was mentioned in section 1.5 that we find additional symmetry in crystals that cannot occur in an isolated molecule. The additional elements differ from those we discussed in Chapter I in that they involve translation as well as rotation or reflexion. Clearly elements involving translation are not possible in a finite molecule, but in a crystal, which is effectively infinite, they frequently occur. Special absences arise from these additional symmetry elements, and before we can understand the absences we must discuss these symmetry elements of which there are only two types, *screw axes* and *glide planes*.

Screw axes, n_p. In this notation, which is universal, n is the order of the axis and p/n is the fraction of the unit cell over which translation occurs. n can have only the values that a pure rotation axis may possess in the solid state, i.e. $n = 2, 3, 4,$ or 6. p can have the values $1, 2, 3,..., (n-1)$. Thus the only possible screw axis based on a rotation axis of order two is a 2_1 axis which must involve rotation through $2\pi/2 = 180°$ and translation by $1/2$ or half the unit cell length. A *twofold screw axis* 2_1 is illustrated in Fig. 1. The coin A represents some part of the contents of the unit cell. The screw axis rotates A through $180°$ about $[a]$ and then moves it through one-half of the cell length in the direction of the screw axis, giving the coin B lying on the opposite side of the paper from A. A similar operation performed on B gives A' which is the point corresponding to A in the next unit cell. A 3_1 axis involves rotation through $2\pi/3 = 120°$ followed by translation of one-third of the unit cell length. The various possible

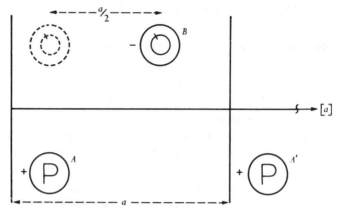

FIG. 1. The operation of a twofold screw axis.

screw axes are drawn out in Fig. 2, in which they are compared with the corresponding pure rotation axes. It will be seen that some of these screw axes (e.g. 3_1 and 3_2, or 4_1 and 4_3) differ only in that they are enantiomorphous.

Glide planes. One possible sort of glide plane is shown in Fig. 3. The starting-point A is reflected across the vertical plane containing $[a]$ and then translated half of the unit cell length to B which must lie on the same side of the paper as A. A similar operation performed on B gives A' in the next unit cell. Glide planes are labelled a, b, or c when the glide is along $[a]$, $[b]$, or $[c]$ respectively. There are two other possible sorts of glide planes. Those labelled n have a translation of $(a+b)/2$, or $(b+c)/2$, or $(a+c)/2$. Those labelled d have a translation of $(a+b)/4$, or $(b+c)/4$, or $(a+c)/4$.

7.2. Systematic absences due to symmetry elements involving translation

The shape of a crystal cannot possibly indicate the existence of symmetry elements involving translation, but the presence of

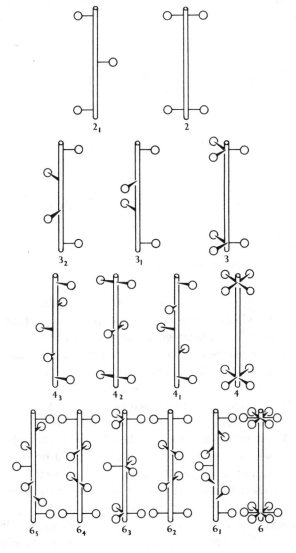

Fig. 2. Comparison of pure rotation axes and screw axes. (Adapted from C. W. Bunn, *Chemical Crystallography*, Clarendon Press.)

screw axes and glide planes can readily be detected by means of X-rays. If we have a crystal that possesses twofold screw axes parallel to [a] (Fig. 4), there is a sheet of atoms exactly half-way between the (100) planes. Thus, as we saw in section 6:10, the

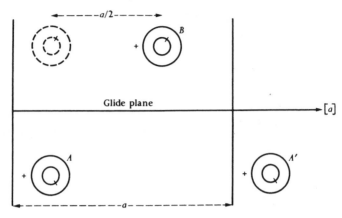

FIG. 3. The operation of a glide plane.

(100) reflexion, and indeed any (h00) reflexion for which h is odd, will be absent. For all other crystal planes there is no simple phase relation between waves scattered from the original and from the interleaved layers of atoms. Consequently the only effect of a twofold screw axis is a halving of the axial reflexions. The extension to screw axes of higher order is quite straightforward and is summarized in Table XXVIII.

A glide plane has a much more marked effect than a screw axis on the X-ray reflexions, because the effect of a glide plane is to halve the whole area of a projection. Thus if a crystal has a c glide perpendicular to [b], the unit cell in projection down [b] appears as though it had a c spacing of only half the true length, as can be seen in Fig. 5, which shows the (h0l) projection of ethylenethiourea (Fig. 7 and section 7.3). Hence, of the (h0l) reflexions, only those with l even will appear. If the glide had been parallel to [a], the a spacing would have appeared halved and the only (h0l) reflexions to appear would have been those

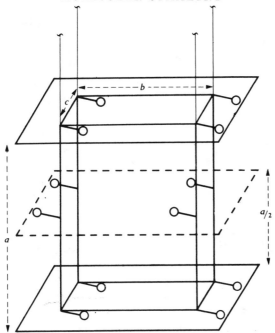

Fig. 4. Effect of twofold screw axis on X-ray reflexions. (Adapted from C. W. Bunn, *Chemical Crystallography*, Clarendon Press.)

TABLE XXVIII

Special absences caused by symmetry elements involving translation†

Symmetry element	Symbol	Special absences
Glide plane ⊥ [b], translation $a/2$.	a	$h0l$ with h odd
Glide plane ⊥ [b], translation $(a+c)/2$	n	$h0l$ with $(h+l)$ odd
Glide plane ⊥ [b], translation $(a+c)/4$	d	$h0l$ with $(h+l)$ not divisible by four
Twofold screw axis ‖ [b] . .	2_1	$0k0$ with k odd
Threefold screw axis ‖ [c] . .	$3_1, 3_2$	$00l$ with l not divisible by three
Fourfold screw axis ‖ [c] . .	$4_1, 4_3$	$00l$ with l not divisible by four
Fourfold screw axis ‖ [c] . .	4_2	$00l$ with l odd
Sixfold screw axis ‖ [c] . .	$6_1, 6_5$	$00l$ with l not divisible by six
Sixfold screw axis ‖ [c] . .	$6_2, 6_4$	$00l$ with l not divisible by three
Sixfold screw axis ‖ [c] . .	6_3	$00l$ with l odd

† Adapted from C. W. Bunn, *Chemical Crystallography*.

with h even. The special absences due to the other sorts of glide planes are shown in Table XXVIII.

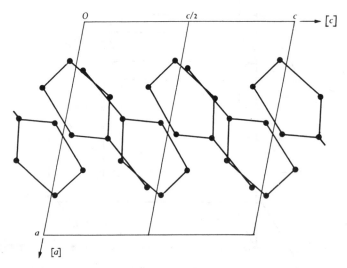

FIG. 5. Projection of the contents of the unit cell of ethylenethiourea down [b]. The pattern repeats twice within the c spacing.

7.3. The space groups

We saw in section 1.5 that there are, in the solid state, only thirty-two point groups. In other words, if we restrict ourselves to rotation and inversion axes of order 2, 3, 4, or 6, there are only thirty-two possible ways of combining symmetry elements. However, this figure does not take into account symmetry elements involving translation. If we include screw axes and glide planes, we find that there is a total of 230 different combinations of symmetry elements possible in the crystalline state. These are known as the 230 *space groups*. They are distributed amongst the seven crystal systems as shown in Table XXIX. Some of the space groups are seldom if ever found in actual crystals, and more than half of the many substances that have so far been investigated belong to the thirteen space groups of the monoclinic system.

TABLE XXIX

The distribution of the space groups amongst the seven crystal systems

Crystal system	Number of space groups
Triclinic	2
Monoclinic	13
Orthorhombic	59
Trigonal	25
Hexagonal	27
Tetragonal	68
Cubic	36

In order to describe the space groups a notation based on the Schoenflies symbols does exist but it is rather arbitrary and uninformative. The Hermann–Mauguin notation is now almost invariably used, and we shall not quote the corresponding Schoenflies symbol. The Hermann–Mauguin symbol starts with a capital letter which refers to the Bravais lattice. Next comes the symbol for the principal axis, whether rotation, inversion, or screw. Thus $P2_1$ indicates a primitive lattice with a twofold axis as the only element of symmetry. $C2$ indicates a lattice centred on the pair of opposite faces defined by $[a]$ and $[b]$, together with a twofold axis. Twofold screw axes are also present in this space group but, as they are implied by the face-centring and the rotation axis, it is not necessary to mention them specifically. If there is a mirror plane or a glide plane perpendicular to the principal axis, the symbol would be written $P2_1/m$ or $C2/c$. Next follow symbols for other axes of equal or lower symmetry and planes of symmetry or glide planes perpendicular to these secondary axes. Thus $P2/m\,2/m\,2/m$ is an orthorhombic space group with three mutually perpendicular twofold axes, and at right angles to each axis a mirror plane. Since the mirror planes imply the presence of the twofold axes, the abbreviated form $Pmmm$ is sufficient and is always used. Finally, in the orthorhombic system, in which the orientation of the axes is arbitrary, the convention is used that the first symbol refers to $[a]$, the second to $[b]$, and the third to $[c]$. Thus the space group symbol $Pbcm$ which, in full, would be

$P2/b\,2_1/c\,2_1/m$ means that, in a primitive orthorhombic lattice, there is a glide plane perpendicular to $[a]$ with a glide of $b/2$, a glide plane perpendicular to $[b]$ with a glide $c/2$, and a mirror plane perpendicular to $[c]$.

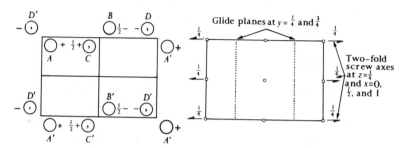

FIG. 6. Diagrams for the space group $P2_1/c$. (Adapted from *International Tables for X-ray Crystallography*, Kynoch Press.)

Diagrams giving the location of the symmetry elements of all the space groups, as well as a great deal of other information invaluable to crystallographers, can be found in the *International Tables for X-ray Crystallography*. In Fig. 6 we have reproduced the diagram for the space group $P2_1/c$, which is the most frequently occurring of all space groups. For clarity each space group diagram is divided into two parts, the right-hand part giving the location of the symmetry elements, and the left-hand part the equivalent positions to which these symmetry elements give rise. The origin is taken at the upper left-hand corner. The x- and y-axes lie in the plane of the paper with the y-axis pointing to the right. The z-axis points upwards from the plane of the paper. Let us select the circle marked A as a starting-point and assign to it the coordinates (x, y, z). The operation of the twofold screw axis, which for convenience is taken to lie at $x = 0$, $z = \frac{1}{4}$, as shown in the right-hand diagram, converts A into B with coordinates $(\bar{x}, \frac{1}{2}+y, \frac{1}{2}-z)$. The bar over a coordinate means that it is negative. The action of the glide plane at $y = \frac{1}{4}$, marked by a dotted line in the right-hand diagram, will convert A into C with coordinates $(x, \frac{1}{2}-y, \frac{1}{2}+z)$, and the glide plane at $y = \frac{3}{4}$ will convert B into D with coordinates $(\bar{x}, 1-y, \bar{z})$. Since C and D were obtained from A and B by reflexion, C and D will be mirror images of A and B and the commas are inserted to show this relationship. Other circles related to A, B, C, and D by translation are marked with primes. A glide plane at right angles to a screw axis always produces a centre of symmetry. The centres of symmetry are marked by small circles in the right-hand diagram. It will be seen that every circle marked A is related to every circle marked D by a centre of symmetry. Similarly, all the circles B and C are centrosymmetrically related. It is always convenient, if possible, to choose a centre of symmetry for the origin. This may be done for the space group $P2_1/c$, and it is this choice that fixes the glide planes and screw axes one-quarter of the way along the axes.

From the above discussion and from Fig. 6 we see that, in a unit cell whose contents are arranged according to the space group $P2_1/c$, there are four equivalent positions whose inter-relation is determined by the symmetry elements. We say that the *multiplicity* of the space group is four. Except for the space group $P1$, which has no symmetry, and the space groups with a single threefold or threefold screw axis, the multiplicity must be an even number and may be as high as 192 for some of the most symmetrical space groups.

Let us now consider an actual example. Crystals of ethylene-thiourea $C_3H_6N_2S$ (Fig. 7) are monoclinic.† There are no general absences amongst the (hkl) reflexions so that the lattice is primitive. The special absences $(0k0)$ for k odd and $(h0l)$ for l odd indicate the presence of a twofold screw axis at right angles to a c glide. The only possible space group is, therefore, $P2_1/c$. The density obtained by flotation and the volume of the unit cell show that there are four molecules in the unit cell. Since there are four general positions in the space group, each molecule must lie in a general position. In other words, the asymmetric unit is one complete molecule. This does not mean that the molecule does not possess symmetry. Fig. 7 shows that the symmetry of the molecule is in fact mm (C_{2v}). What it does mean is that the molecules can form a system of lower potential energy

† P. J. Wheatley, *Acta crystallogr.* 1953, **6**, 369.

if they pack into a unit cell without using such symmetry as they do possess. Consequently a determination of the space

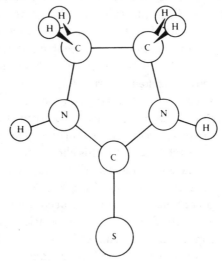

FIG. 7. The molecule of ethylenethiourea.

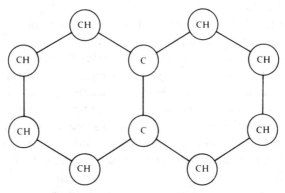

FIG. 8. The molecule of naphthalene.

group of ethylenethiourea cannot tell us anything about the molecular symmetry.

Now we consider the naphthalene molecule (Fig. 8). Naphthalene $C_{10}H_8$ also crystallizes in the space group $P2_1/c$, but a density determination shows that there are only two molecules in the unit cell.† In this case the asymmetric unit cannot be the whole molecule but must be half the molecule, and the two halves must be related by one of the symmetry elements present in the space group. The only symmetry elements possessed by the space group $P2_1/c$ are a twofold screw axis, a glide plane, and a centre of symmetry. The two halves of a finite molecule cannot be related by a screw axis or a glide plane, since these symmetry elements involve translation, so we may conclude that the naphthalene molecule must possess a centre of symmetry. The molecules must be located in the unit cell in such a way that the centre of symmetry of a molecule coincides with a centre of symmetry in the unit cell. The molecule is said to lie in a *special position*, and whenever this occurs the molecule must possess symmetry of some sort. The determination of the space group of naphthalene has not told us the full symmetry of the molecule, which is mmm (D_{2h}), but it has given us a useful start.

A crystallographer always hopes that his unit cell will contain fewer molecules than there are general positions, because he will then know that his asymmetric unit is not the whole molecule but only part of it. In section 7.5 we shall give some more examples of the way in which the determination of a space group can help in the elucidation of molecular symmetry, but first we must say a little more about the determination of space groups.

7.4. Determination of space groups

The number of possible space groups for a crystal under investigation is almost always limited by the fact that we have already determined the crystal system either by looking at the external shape of the crystal or by measuring the size and shape of the unit cell. The Bravais lattice is determined uniquely by the general absences, and this reduces the number of possible space groups still further. A final choice can often, but not always, be made from the special absences. Thus, as we saw in the last section, the space group $P2_1/c$ can be determined

† S. C. Abrahams, J. M. Robertson, and J. G. White, *Acta crystallogr.* 1949, **2**, 233, 238.

unambiguously from the special absences amongst the $(0k0)$ and $(h0l)$ reflexions. Similarly, if the only absences are $(h00)$ for h odd, $(0k0)$ for k odd, and $(00l)$ for l odd, the space group must be the orthorhombic $P2_12_12_1$ with three twofold screw axes at right angles. However, X-rays cannot detect directly the presence of certain symmetry elements such as a centre of symmetry or a mirror plane, since these elements possess no translational component and thus do not give rise to special absences. Some space groups differ from others only by the presence of these symmetric elements and then X-rays cannot make the distinction.

In these circumstances it may be possible to confirm the absence of a centre of symmetry by positive piezoelectric or pyroelectric effects.† Sometimes the external shape of the crystal proves to be helpful. More importantly it has been shown that it is possible to detect a centre of symmetry by a consideration, not of the absent reflexions, but of the distribution of intensity amongst the various X-ray reflexions that are present. In most cases a combination of these various methods will decide the space group, but occasionally it is necessary to find the actual positions of all the atoms by the methods described in the next chapter and to show that these positions are consistent with the symmetry of one particular space group rather than any other.

7.5. Examples of the use of space groups for the determination of molecular symmetry

As in the case of naphthalene, discussed in section 7.3, it is seldom that a determination of the space group gives the full molecular symmetry. Nevertheless, the information obtained is often sufficient, when combined with chemical knowledge, to leave little doubt as to the correct structure. We shall consider a number of examples to show the extent of the information that can be gained.

Urea $(NH_2)_2CO$.‡ Urea forms tetragonal crystals. A determination of the unit-cell dimensions and of the density shows

---

† For a discussion of piezoelectricity and pyroelectricity see W. A. Wooster, *Crystal Physics* (Cambridge University Press).

‡ P. A. Vaughan and J. Donohue, *Acta crystallogr.* 1952, **5**, 530.

that there are only two molecules in the unit cell. There are no general absences amongst the X-ray reflexions, and the only special absences are the (h00) reflexions with h odd and, since in a tetragonal crystal [a] and [b] are equivalent, the ($0k$0) reflexions with k odd. Reference to *International Tables* shows that these absences give $P\bar{4}2_1m$ and $P42_12$ as the possible space groups. We can immediately exclude the latter since it would require the urea molecule to have either a fourfold axis or three

TABLE XXX
Equivalent positions for the space group $P\bar{4}2_1m$

Multiplicity of positions	Point symmetry	Coordinates of equivalent positions
8	1 (C_1)	x, y, z; $\frac{1}{2}-x, \frac{1}{2}+y, \bar{z}$; $\bar{x}, \bar{y}, z$; $\frac{1}{2}+x, \frac{1}{2}-y, \bar{z}$; $\bar{y}, x, \bar{z}$; $\frac{1}{2}+y, \frac{1}{2}+x, z$; $y, \bar{x}, \bar{z}$; $\frac{1}{2}-y, \frac{1}{2}-x, z$.
4	m (C_h)	$x, \frac{1}{2}+x, z$; $\bar{x}, \frac{1}{2}-x, z$; $\frac{1}{2}+x, \bar{x}, \bar{z}$; $\frac{1}{2}-x, x, \bar{z}$.
4	2 (C_2)	$0, 0, z$; $0, 0, \bar{z}$; $\frac{1}{2}, \frac{1}{2}, z$; $\frac{1}{2}, \frac{1}{2}, \bar{z}$.
2	mm (C_{2v})	$0, \frac{1}{2}, z$; $\frac{1}{2}, 0, \bar{z}$.
2	$\bar{4}$ (S_4)	$0, 0, \frac{1}{2}$; $\frac{1}{2}, \frac{1}{2}, \frac{1}{2}$.
2	$\bar{4}$ (S_4)	$0, 0, 0$; $\frac{1}{2}, \frac{1}{2}, 0$.

intersecting twofold axes. This is clearly impossible so we may concentrate on the space group $P\bar{4}2_1m$. Part of the information for this space group is reproduced from *International Tables* in Table XXX. It will be seen that there are eight general positions in which the whole molecule is the asymmetric unit, and for which no molecular symmetry is required. There are two sets of positions with fourfold multiplicity, and three with twofold multiplicity. Since we know that the unit cell of urea contains only two molecules, we must be concerned with one of these last three sets. Two of these sets require the molecule to have $\bar{4}$ (S_4) symmetry, which is clearly impossible with a molecule of the formula $(NH_2)_2CO$. The other set requires the molecule to have mm (C_{2v}) symmetry, which is quite feasible. In fact it leaves only two possible molecular configurations, differing in the positions of the hydrogen atoms. In one the whole molecule is planar

(Fig. 9), and in the other the planes containing the hydrogen atoms are at right angles to the plane containing the carbon, oxygen, and nitrogen atoms (Fig. 10). The complete molecular geometry of the urea molecule may be found by various methods and we shall see in section 13.7 how nuclear magnetic resonance studies can decide which of these two possible configurations of mm (C_{2v}) symmetry is the correct one.

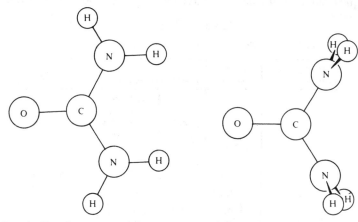

FIG. 9. The planar form of the urea molecule.

FIG. 10. The non-planar form of the urea molecule. The symmetry, like that of the planar form, is mm (C_{2v}).

Silver azide AgN$_3$.† Silver azide forms orthorhombic crystals. The general absences of (hkl) reflexions with ($h+k+l$) odd show that the crystal structure is based on a body-centred lattice. The special absences ($0kl$) with k odd and ($h0l$) with h odd show that the space group is either *Iba* or *Ibam*, which differ in that the latter has an additional mirror plane perpendicular to [c]. *Iba* has eight general positions whereas *Ibam* has sixteen. There are four formula units in the cell. The absence of any detectable pyroelectric effect suggests, but does not prove, that the centrosymmetrical space group *Ibam* is the more likely. In this space group there are four sets of positions with fourfold multiplicity. Two of these sets require the symmetry of the azide ion

† M. Bassière, *C.r. hebd. Séanc. Acad. Sci., Paris* 1935, **201**, 735.

to be $2/m$ (C_{2h}), and the other two require the symmetry to be
222 (D_2). In either case the azide ion must be linear and symmetrical with ∞/mm ($D_{\infty h}$) symmetry. If, however, the space
group is Iba, the required point symmetry is 2 (C_2). In this case
the only information obtained is that the two bonds must be of
the same length or the ion must be linear, depending on whether
the twofold axis bisects the NNN angle or whether it coincides
with the lines joining the three nitrogen atoms. The correct
space group was decided from a complete structure analysis
which showed that the atoms were arranged in accordance with
the space group $Ibam$. Consequently the ion must be linear
and symmetrical.

Tetraboron tetrachloride B_4Cl_4.† The reactive compound B_4Cl_4
forms tetragonal crystals. No general absences are observed but

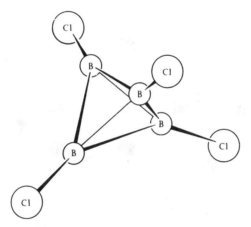

FIG. 11. The molecule of tetraboron tetrachloride, B_4Cl_4.

reflexions (hhl) with l odd are missing as well as ($hk0$) with
($h+k$) odd. These absences lead unambiguously to the space
group $P4_2/nmc$. The reactive nature of the crystals preclude
direct observation of the density, but a reasonable calculated
density is obtained on the assumption that there are two molecules in the unit cell. The space group has sixteen general
positions. There are two sets of positions with twofold multi-

† M. Atoji and W. N. Lipscomb, *Acta crystallogr.* 1953, **6**, 547.

plicity and both sets require the point symmetry of the molecule to be $\bar{4}2m$ (D_{2d}). The only reasonable chemical formula is one in which a tetrahedron of boron atoms is surrounded by a larger tetrahedron of chlorine atoms as shown in Fig. 11. If both tetrahedra were regular, the symmetry of the molecule would be $\bar{4}3m$ (T_d), and a complete X-ray analysis shows that the molecule differs only slightly from this higher symmetry.

s-Triazine (CHN)₃.† Stoichiometrically *s*-triazine (Fig. 12) is the trimer of hydrogen cyanide. Many years ago HCN was polymerized in the presence of HCl and the molecular weight of the product determined cryoscopically. The formula (HCN)₂

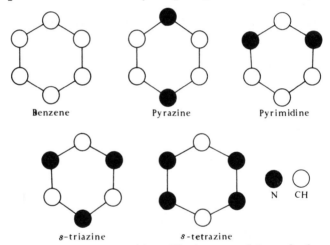

FIG. 12. Benzene and the azines. The departure of the angles from 120° is exaggerated so as to emphasize the symmetry of each molecule.

was assigned as a result of this investigation. More recently this work was repeated as a preliminary study to the synthesis of *s*-triazine, which had remained undetected. The later investigation‡ showed that the correct formula for the polymerization product was in fact (HCN)₃, and it remained to be discovered whether the product was *s*-triazine. The physical properties, especially the high volatility, were consistent with a highly symmetrical molecule. Optical examination of the crystals

† P. J. Wheatley, *Acta crystallogr.* 1955, **8**, 224.
‡ C. Grundmann and A. Kreutzberger, *J. Am. chem. Soc.* 1954, **76**, 632.

showed them to be trigonal. X-ray photographs confirmed the threefold symmetry and yielded the dimensions of the unit cell. There were no general absences, but reflexions of the type (hhl) with l odd were absent. These special absences characterize the space group as $R\bar{3}c$. A density determination showed that there were only two molecules in the unit cell, whereas the space group has twelve general positions. There are two sets of positions with a multiplicity of two. One set requires the point symmetry of the molecule to be $\bar{3}$ (S_6). This cannot possibly be the symmetry of $(HCN)_3$ since this point group has a centre of symmetry, and there is no way in which three hydrogen atoms, three carbon atoms, and three nitrogen atoms can be arranged centrosymmetrically. The other set requires the point symmetry to be 32 (D_3) with three twofold axes at right angles to a threefold axis. Although there are other ways of joining these atoms together with this symmetry, the only chemically reasonable way is that shown in Fig. 12.

Benzene and several of the azines have been investigated by X-ray diffraction and in all cases but one the space group has given some information about the molecular symmetry. The different molecules are shown in Fig. 12, and the crystallographic results are listed in Table XXXI. The fifth column of this Table gives the *crystallographic symmetry*, that is, the minimum symmetry that the molecule must possess according to the space-group requirements.

TABLE XXXI

Crystallographic information for benzene and some azines

Molecule	Space group	No. of molecules in unit cell	Multiplicity	Cryst. symmetry	True symmetry
Benzene† .	$Pbca$	4	8	$\bar{1}$ (S_2)	$6/mmm$ (D_{6h})
Pyrazine‡ .	$Pmnn$	2	8	$2/m$ (C_{2h})	mmm (D_{2h})
Pyrimidine§ .	Pna	4	4	None	mm (C_{2v})
s-Triazine‖ .	$R\bar{3}c$	2	12	32 (D_3)	$\bar{6}2m$ (D_{3h})
s-Tetrazine¶ .	$P2_1/c$	2	4	$\bar{1}$ (S_2)	mmm (D_{2h})

† E. G. Cox, D. W. J. Cruickshank, and J. A. S. Smith, *Proc. R. Soc.* A 1958, **247**, 1.   ‡ P. J. Wheatley, *Acta crystallogr.* 1957, **10**, 182.
§ P. J. Wheatley, ibid. 1960, **13**, 80.
‖ P. J. Wheatley, ibid. 1955, **8**, 224.
¶ F. Bertinotti, G. Giacomello, and A. M. Liquori, ibid. 1956, **9**, 510.

140 X-RAY DIFFRACTION

Apart from the many substances for which a knowledge of the space group gives no information about the molecular symmetry, there are a few known for which the space group requirements cannot possibly represent the true symmetry of the molecule. The molecules apparently have higher symmetry than expected. These anomalies are due either to the fact that there is a statistical distribution of less symmetrical molecules within the unit cells, or to the fact that the molecules are rotating in the crystal. For example, the compound $[Co(NH_3)_3H_2OCl_2]Cl$, whose cation is shown in Fig. 13, crystallizes in the space group $P6_3/mmc$ with two formula units in the cell.[†] The cations are in special positions that require them to have $\bar{6}2m$ (D_{3h}) symmetry. This is clearly impossible, and it is found that the crystal actually consists of a statistical array of the less symmetrical complexes arranged in such a way that, on a macroscopic scale, they appear to have the higher symmetry.

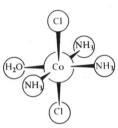

FIG. 13. The ion
$[Co(NH_3)_3(H_2O)Cl_2]^+$.

An example of rotation in the solid state is provided by the complex formed between silver perchlorate and dioxane, $AgClO_4.3C_4H_8O_2$.[‡] This compound crystallizes in a primitive cubic cell containing one formula unit. The probable space group is $Pm3m$ which has two sites of onefold multiplicity, each site having point symmetry $m3m$ (O_h). The silver ion can be placed on one of these two sites, but it is inconceivable that the whole of the rest of the complex can have octahedral symmetry. It is found that the crystal actually consists of silver ions at the corners of the cube, each surrounded by a regular octahedron of dioxane oxygen atoms. The dioxane molecules lie along the principal axes of the cube and are rotating. The perchlorate ion is situated on the other site of $m3m$ (O_h) symmetry at the body centre of the cube, and this ion is also rotating. Fig. 14 shows one face of the unit cube with the dioxane molecules represented in various rotational orientations.

[†] Y. Tanito, Y. Saito, and H. Kuroya, *Bull. chem. Soc. Japan* 1952, **25**, 328.
[‡] R. J. Prosen and K. N. Trueblood, *Acta crystallogr.* 1956, **9**, 741.

These anomalous cases are rare, and they are usually so striking that there is little likelihood of an incorrect allocation of symmetry to the molecule.

7.6. Determination of bond lengths from cell dimensions

So far we have been concerned only with the positions of the X-ray reflexions, and with whether or not some particular reflexions are absent. This information is sufficient to give us the size and shape of the unit cell and, in some cases, to tell us

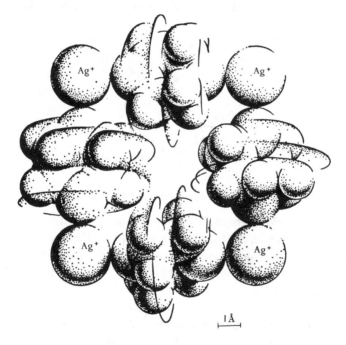

Fig. 14. A section at $x = 0$ through the crystal of $AgClO_4 . 3C_4H_8O_2$. The dioxane molecules are shown in various rotational orientations. The perchlorate ion, which is located at $\frac{1}{2}\frac{1}{2}\frac{1}{2}$, extends to within about 1 Å of the plane $x = 0$.

some, if not all, of the symmetry elements possessed by the molecule. The positions of the X-ray reflexions can be measured with great precision, and cell dimensions can be obtained with an accuracy of ± 0.001 Å without too much difficulty. With care the cell side of a cubic crystal at a fixed temperature can be obtained to ± 0.00001 Å. In the examples quoted in the previous section, the positions of the X-ray reflexions gave us no information about the actual bond lengths in the molecules. However, in the case of sodium chloride (section 6.11) it is clear that the nearest distance between a sodium and a chloride ion is equal to half the cell side. Consequently this interionic distance can be obtained very accurately. The same situation holds for many ionic crystals and for most metals, but it is unusual to find a covalent distance that is a function only of the cell side. The

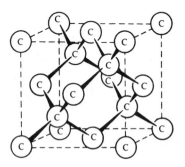

FIG. 15. The crystal structure of diamond.

only known cases are the cubic modifications of the Group IV elements, of which the best-known example is diamond (Fig. 15). In this crystal, which is really a giant molecule, each atom must, by symmetry, be joined tetrahedrally to the nearest four neighbours. It can easily be seen that the length of the C—C bond is given by $a\sqrt{3}/4$. A careful determination of the unit cell dimension at 20° C gave a value $a = 3.56679$ Å. Consequently $r_0(CC)$ must be 1.54447 Å at this temperature.[†]

We shall discuss in the next chapter the steps that normally have to be taken in order to obtain molecular parameters.

† M. E. Straumanis and E. Z. Aka, *J. Am. chem. Soc.* 1951, **73**, 5643.

VIII

X-RAY DIFFRACTION: DETERMINATION
OF MOLECULAR PARAMETERS

8.1. Introduction

W E have seen how, in certain cases, the symmetry of a molecule in a crystal can be established from the positions of the X-ray reflexions and from the systematic absences. In general, of course, this is insufficient to establish the molecular parameters, and in some way we have to be able to locate all the atoms within the unit cell. In order to do this we have to take into account the actual intensity of the observed X-ray reflexions.

X-rays are scattered almost entirely by the extranuclear electrons of an atom, and the intensity of the scattered radiation must depend on the distribution within the atom of all the electrons. For small angles of diffraction the amplitude of the scattered beam is equal to the sum of the amplitudes of the beams that would be scattered by each electron individually. In these circumstances the total amplitude is proportional to the number of extranuclear electrons. For atoms this is equal to the atomic number Z, but for ions the number of extranuclear electrons differs from Z by the charge on the ion. At higher angles of diffraction the different scattered rays will interfere, the scattering will be reduced, and the proportionality factor will become less than the number of extranuclear electrons. This proportionality factor is known as the *atomic scattering factor f.* Scattering factors can be calculated from the wave-functions of the electrons, and the results have been tabulated. Some examples of scattering factors are plotted as a function of $\sin\theta/\lambda$ in Fig. 1. As usual θ is the Bragg angle and λ is the wavelength of the X-rays. Electronic wave-functions are constantly being refined and new scattering factors calculated. However, the modifications are slight and are insignificant for our present purposes.

If X-rays are scattered by fluids, a radially symmetric diffraction pattern is produced just like an electron diffraction pattern (section 5.1). The scattering of X-rays is much less than that of electrons, however, and exposures of several hours are required

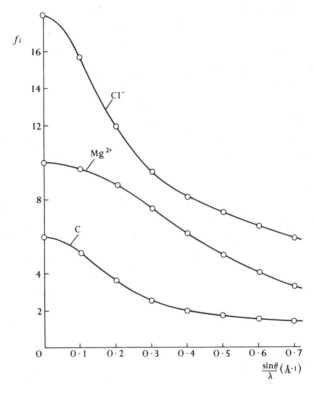

FIG. 1. Atomic scattering factor curves for C, Mg^{2+}, and Cl^-.

whereas a few seconds is sufficient for electrons. When the scattering molecules are arranged regularly in a crystal instead of randomly in a gas or liquid, the scattering is of a very different kind, as we saw in section 6.7, and depends on the arrangement of the molecules in the unit cell. One set of planes may have a high density of atoms, in which case diffraction from this set will be considerable, and an intense diffracted beam will be produced. Another set of planes may be only sparsely populated, in which case a diffracted beam of weak intensity will be obtained. The problem in crystallography is to relate each observed intensity from a known set of planes (*hkl*) to the distribution of atoms within this set of planes. If this can be done for many sets of planes within the crystal, the distribution of the atoms within the unit cell is automatically known.

The principle of the method can be stated quite simply. For each set of planes (*hkl*) we measure the intensity I_{hkl} of the diffracted beam. We have already seen in section 6.9 how the X-ray reflexions can be indexed, and we shall discuss in the next section methods for the measurement of intensities. For crystals consisting of a large number of crystallites (section 6.3) this intensity, apart from a geometrical factor that can be calculated, is equal to the square of a quantity which we call the *structure factor* F_{hkl}. When this geometrical factor has been applied, we shall refer to the intensities as the *corrected intensities*. Thus from each observed intensity we can obtain a corrected intensity whose square root gives the observed structure factor F_{hkl}^{obs}. Now we can also calculate a structure factor, since it is a function of only two properties of the atoms within the unit cell, their scattering factors and their positions. Consequently, if there are N atoms in the unit cell, we may write

$$F_{hkl}^{calc} = \sum_{i=1}^{N} f_i \times Q(x_i, y_i, z_i),$$

where F_{hkl}^{calc} is the calculated structure factor for the set of planes (*hkl*) and $Q(x_i, y_i, z_i)$, known as the *geometric structure factor*, is a function of the coordinates x_i, y_i, and z_i of the ith atom. We already know values of f_i, as explained earlier in this section, so that all we have to do is to postulate positions for the atoms and calculate values of F_{hkl}^{calc}. If we have guessed the correct positions, the various calculated structure factors will agree with the corresponding observed structure factors. If the selected atomic coordinates are not correct, agreement between the observed and calculated structure factors will be lacking, and the atoms must be moved so as to improve the agreement.

The fact that there is an infinite number of possible positions for each atom in the unit cell apparently makes this trial-and-error method impracticably long. We shall see that this is not so because there are various pieces of information that we can use to define the atomic positions more closely. On the other hand, there are various points that have been glossed over in the above description, and we must now examine them in more detail in order to understand some of the difficulties that surround a crystal structure determination.

8.2. Measurement of intensities

The measurement of an intensity is one of the most inaccurate and least satisfactory of scientific operations. Two methods are used in X-ray crystallography, and each has some advantages over the other. In the visual method the diffracted beams are allowed to fall on a photographic film for many hours, and the relative intensities of the spots are estimated by visual comparison with a standard strip. The strip is made by exposure of a film to a diffracted beam, preferably from the same crystal, for different known intervals of time. A wide range of intensities may be encompassed by use of the *multiple-film technique* in which a pack of three or four films placed one behind the other is exposed to the diffracted X-rays. The intensity falls off by a factor of about 3 from one film to the next. The eye is a good integrator, and, with a little experience, the whole of the blackening within a spot can be estimated to within ± 20 per cent. This means that the random error in an observed structure factor may be ± 10 per cent. In favourable cases these figures may be lowered quite appreciably. The two main advantages of the photographic method are firstly that many reflexions are recorded on the same film, and secondly, since exposures are of long duration, any changes in the operating conditions are smoothed out.

It may seem that these are rather crude observations on which to base a structure analysis, and indeed they are. But X-ray diffraction differs from most other methods used for structure determination in that there is almost always a considerable surfeit of experimental observations. A crystal made up of molecules containing ten atoms may be expected to give of the order of a thousand independent reflexions with X-rays of a wavelength of 1·5 Å. Since only thirty coordinates are needed to define the positions of the ten atoms, we have a large number of redundant observations. However, when observations suffer from a random error, the error in a derived result is proportional to $\{1/(N-p)\}^{\frac{1}{2}}$, where N is the number of observations and p the number of parameters to be determined. Clearly, then, we need to make as many observations as we can in order to reduce errors in the atomic coordinates. This combination of relatively large random errors and a surfeit of observations distinguishes X-ray diffraction from most other methods discussed in this book. The large number of observations means that we can deal with large molecules, but the random errors mean that we can locate the atoms with only a limited accuracy.

In order to increase the accuracy of photographic measurements, integrating photometers have been used to measure the intensities of the spots, but the errors are still of the order of ± 10 per cent in the intensities. Much more accurate intensities may in principle be obtained with a spectrometer equipped with a Geiger counter or a scintillation counter. In this way errors can, in favourable circumstances, be reduced to ± 3 per cent. However, since counter methods measure each reflexion separately, they are much more susceptible to fluctuations or changes in the operating conditions.

Apart from these random errors, X-ray intensities also suffer from systematic errors caused by such factors as absorption of the diffracted beam as it passes through the crystal. Some of these systematic errors can be eliminated by a careful choice of crystal, or may be allowed for in subsequent calculations, but it is frequently not possible to choose an ideal crystal and the necessary corrections are often ignored or made only imperfectly. Consequently, additional inaccuracies arising from these systematic errors are almost always present in the observed intensities. This is an unsatisfactory state of affairs, and one that must be corrected, but in the meantime we must accept this limitation and realize that X-ray diffraction will continue to yield less accurate molecular parameters than most other experimental methods. Since errors are present in the observed structure factors, we can never make our calculated and observed structure factors agree, and we must have some criterion to tell us whether the measure of agreement is good enough or not. We shall see in section 8.7 what criterion we employ.

8.3. The geometric structure factor

In section 8.1 we described the geometric structure factor $Q(x_i, y_i, z_i)$ as some function of the atomic coordinates of the ith atom. We must now define this function more precisely. When we discussed cubic crystals in section 6.10, we mentioned only very simple crystals in which a single atom was placed at the corners, body-centre, or face-centres of the cube. Suppose we replace the atom by an N-atomic molecule, and suppose that the ith atom in the molecule is located at the point x_i, y_i, z_i. If a beam of X-rays is reflected from a plane (hkl) passing through this point, and also from a parallel plane passing through the origin, there will be a phase difference between the two reflected waves given by

$$\phi_i = 2\pi(hx_i/a + ky_i/b + lz_i/c). \qquad (1)$$

x/a, y/b, and z/c are called the *fractional coordinates* of an atom. For each atom in the molecule there will be an expression similar to equation (1), and these expressions will differ only in that each one involves the coordinates of a different atom. In order to calculate the intensity of the beam diffracted by the whole molecule, we need to know the amplitude and the phase of the wave diffracted by each atom. We can then compound these individual waves to obtain the total wave. Since equation (1) gives us the phase, and since the amplitude of each wave is proportional to the scattering factor of the atom responsible for the wave, the corrected intensity, which is equal to the square of the total amplitude, is given by

$$I_{hkl} = F_{hkl}^2 = \left(\sum_{i=1}^{N} f_i \cos \phi_i \right)^2 + \left(\sum_{i=1}^{N} f_i \sin \phi_i \right)^2. \qquad (2)$$

If the crystal possesses a centre of symmetry which we select as the origin, the last term in equation (2) vanishes, since for every atom in the structure with a phase ϕ there is another, related to the first by the centre of symmetry, with a phase $-\phi$. In all that follows we shall assume that our crystal does have a centre of symmetry. We can then reduce equation (2) to

$$I_{hkl} = F^2_{hkl} = \Big(\sum_{i=1}^{N} f_i \cos \phi_i \Big)^2. \qquad (3)$$

We may illustrate the application of equation (3) by a consideration of various sorts of face-centred cubic structures.

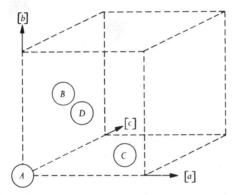

Fig. 2. The four atoms that define a face-centred cubic crystal.

Suppose first of all that our crystal is made up of identical atoms situated at each corner and at each face-centre of the unit cell. As we saw in section 6.5, the unit cell contains only four atoms, and we can choose the four atoms A, B, C, and D shown in Fig. 2. The fractional coordinates of these four atoms are 000, $0\frac{1}{2}\frac{1}{2}$, $\frac{1}{2}0\frac{1}{2}$, and $\frac{1}{2}\frac{1}{2}0$ respectively. Consequently, from equations (1) and (3),

$$I_{hkl} = F^2_{hkl} = f\{\cos 0 + \cos 2\pi(\tfrac{1}{2}k + \tfrac{1}{2}l) + \cos 2\pi(\tfrac{1}{2}h + \tfrac{1}{2}l) +$$
$$+ \cos 2\pi(\tfrac{1}{2}h + \tfrac{1}{2}k)\}^2. \qquad (4)$$

Equation (4) can be rearranged to give

$$I_{hkl} = F^2_{hkl} = 4f\{\cos \pi(h+k+l)\cos \tfrac{1}{2}\pi(h+l)\cos \tfrac{1}{2}\pi(k+l) \times$$
$$\times \cos \tfrac{1}{2}\pi(h+k)\}^2. \qquad (5)$$

Equation (5) gives us the condition for the occurrence of reflexions from different planes of a face-centred cubic crystal. Since $\cos\frac{1}{2}\pi = \cos\frac{3}{2}\pi = \cos\frac{5}{2}\pi = \ldots = 0$, equation (5) will vanish if $(h+l)$ or $(k+l)$ or $(h+k)$ is odd. In other words all three indices must be odd or all three even, which is the result we obtained in section 6.10.

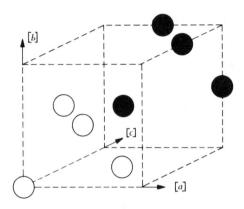

FIG. 3. The four pairs of ions that define a crystal of the type formed by NaCl.

Now suppose we have a face-centred cubic structure of the type formed by sodium chloride (Fig. 6.21). In order to obtain the geometric structure factor of this arrangement of ions, it is best to regard the crystal as made up of a face-centred array of sodium ions and a face-centred array of chloride ions, the two arrays being displaced by half the body diagonal of the unit cube. In other words, for every sodium ion with coordinates x, y, z there is a chloride ion with coordinates $\frac{1}{2}+x, \frac{1}{2}+y, \frac{1}{2}+z$, as shown in Fig. 3. If we use equations (1) and (3) as before, and consider a sodium ion to be at the origin, we obtain for the corrected intensity of the diffracted beam from the set of planes (hkl)

$$I_{hkl} = F^2_{hkl} = f_{\mathrm{Na}}\{\cos 0 + \cos 2\pi(\tfrac{1}{2}k + \tfrac{1}{2}l) +$$
$$+ \cos 2\pi(\tfrac{1}{2}h + \tfrac{1}{2}l) + \cos 2\pi(\tfrac{1}{2}h + \tfrac{1}{2}k)\}^2 + f_{\mathrm{Cl}}\{\cos 2\pi(\tfrac{1}{2}h + \tfrac{1}{2}k + \tfrac{1}{2}l) +$$
$$+ \cos 2\pi(\tfrac{1}{2}h + k + l) + \cos 2\pi(h + \tfrac{1}{2}k + l) + \cos 2\pi(h + k + \tfrac{1}{2}l)\}^2.$$

This equation may be rearranged to give

$$I_{hkl} = F^2_{hkl} = 4f_{Na}\{\cos\pi(h+k+l)\cos\tfrac{1}{2}\pi(h+l)\cos\tfrac{1}{2}\pi(k+l)\times$$
$$\times\cos\tfrac{1}{2}\pi(h+k)\}^2 + (-1)^{(h+k+l)}[4f_{Cl}\{\cos\pi(h+k+l)\cos\tfrac{1}{2}\pi(h+l)\times$$
$$\times\cos\tfrac{1}{2}\pi(k+l)\cos\tfrac{1}{2}\pi(h+k)\}^2]. \qquad (6)$$

Thus, again, for reflexions to appear, h, k, and l must all be odd or all even. Moreover, we see that, if h, k, and l are all even, the two terms on the right-hand side of equation (6) have the same sign and the contributions from the sodium and chloride ions will reinforce, but if h, k, and l are all odd, the signs of these two terms are opposite and the contributions will interfere. Since the scattering factors of sodium and chloride ions are different, all reflexions with h, k, and l all even or all odd will appear, but those with odd values of the indices will be weaker than those with even values. This alternating intensity effect can be seen in Figs. 6.13 and 6.20. In potassium chloride, which has the same crystal structure as sodium chloride, an additional feature arises since the potassium and chloride ions are isoelectronic and have practically the same scattering factors. Consequently, for odd indices the two terms on the right-hand side of equation (6) will almost exactly cancel, and we shall have this extra set of absences. Only reflexions for which h, k, and l are all even will appear at all strongly, those for h, k, and l odd being very weak, and the crystal structure of potassium chloride will appear to be based on a simple cubic lattice with only half the true spacing, unless a very close scrutiny reveals the presence of these weak reflexions.

8.4. Fourier series

In the last section we discussed how, given the atomic coordinates, we can calculate the intensities of all the possible X-ray reflexions. Only in the simplest of crystal structures is it possible to do this directly. In general we do not know the atomic positions, and we have to be able to answer the reverse question: Given the intensities, how can we determine the atomic positions? A method for doing this has been proposed by Bragg, and practically all crystal structure determinations depend on this method or some modification of it.

Any quantity that is a periodic function of a single coordinate x can be represented by a *Fourier series*

$$F(x) = A_0 + A_1 \cos 2\pi(x/a + \alpha_1) + A_2 \cos 2\pi(2x/a + \alpha_2) + \dots$$

$$= \sum_{n=0}^{\infty} A_n \cos 2\pi(nx/a + \alpha_n), \tag{7}$$

where A is the amplitude of each cosine wave, α is the phase angle, and a is the repeat distance. If the function is centrosymmetric, the phase angles can be only $0°$ or $180°$, and equation (7) reduces to the form

$$F(x) = \sum_{n=0}^{\infty} A_n \cos 2\pi nx/a. \tag{8}$$

Each coefficient must then be given the appropriate sign, positive if the phase angle is $0°$ and negative if the phase angle is $180°$.

All that equation (8) means is that any linear centrosymmetric pattern that repeats after a certain distance a can be accurately represented by the superposition of an infinite number of cosine waves of different amplitudes and wavelengths.

The density of scattering matter within a crystal is a periodic function in three dimensions, and we can represent this electron density by a triple Fourier series

$$\rho(xyz) = \sum_{-\infty}^{\infty} \sum_{-\infty}^{\infty} \sum_{-\infty}^{\infty} A_{hkl} \cos 2\pi(hx/a + ky/b + lz/c). \tag{9}$$

It can be shown that the amplitude A_{hkl} is equal to the corresponding structure factor F_{hkl} divided by the volume of the unit cell. Consequently we may write equation (9) in the form

$$\rho(xyz) = \frac{1}{V} \sum_{-\infty}^{\infty} \sum_{-\infty}^{\infty} \sum_{-\infty}^{\infty} F_{hkl} \cos 2\pi(hx/a + ky/b + lz/c). \tag{10}$$

In principle equation (10) permits the complete solution of the structure of any crystal. All we need to know are the cell dimensions and the observed structure factors. The Fourier series may then be evaluated, and will give a picture of the scattering matter within the unit cell. At the atomic positions the electron density $\rho(xyz)$ will rise to a maximum, whereas in between the atoms $\rho(xyz)$ will fall to zero. Strictly speaking we should use an infinite number of structure factors in the series, but in practice we have, of course, to make do with a finite number. This causes some errors in the positions of the atomic peaks, but in X-ray diffraction work the errors are not serious because the fall-off in the atomic scattering curves with increasing Bragg angle ensures that the structure factors get smaller at higher Bragg angles. Consequently we have a converging series which is the next best thing to an infinite one. Moreover, as we shall see in section 8.7, there is a method available for the removal of almost all the finite-series error.

Equation (10) can be simplified to two dimensions or to one; in two dimensions we need to use only one zone of reflexions such as $(hk0)$ and in one dimension only one set of axial reflexions such as $(h00)$. Since we are using fewer observations we shall obtain less information, and a two-dimensional Fourier gives only the contents of the unit cell projected on to a face of the cell. A one-dimensional Fourier synthesis, which is seldom of much value, gives the entire contents of the cell projected on to a single line.

In order to compute the electron density it is necessary to evaluate equation (10) only for selected values of the variables x, y, and z. It is customary to choose $12°$, $6°$, or $3°$ as suitable intervals of $2\pi x/a$, $2\pi y/b$, or $2\pi z/c$, so that values of the electron density are obtained every $\frac{1}{30}$th, $\frac{1}{60}$th, or $\frac{1}{120}$th of the cell side. In this way a net of points is built up which can be joined by lines to form a contour map. The resulting Fourier map then gives a picture of the contents of the unit cell in three dimensions, or projected on to a face, or projected on to a line, depending on whether we have used a triple, double, or single Fourier series. Various numerical aids have been devised for simplifying the lengthy calculations needed in summing a Fourier series. Values of $F_{hkl} \cos 2\pi hx/a$ have been tabulated for values of h from 1 to 20, unit intervals of F_{hkl} up to 100, and $3°$ or $6°$ intervals of $2\pi x/a$. With the help of these tables a two-dimensional Fourier synthesis could be calculated and drawn out in a day. Little work had been done with three-dimensional Fourier series until the advent of electronic computers. The speed of these machines is such that three-dimensional Fourier syntheses can be calculated in a matter of minutes, thus permitting crystallographers to employ the much more powerful three-dimensional methods. We shall give examples of the use of Fourier analysis shortly, but first we must consider one more fundamental difficulty.

8.5. The phase problem

The observed intensity is proportional to the square of the structure factor. Thus, although we know the numerical magnitude of each F_{hkl} to use in equation (10), we do not know the sign. No one has yet devised an experiment that will determine the phase angle of each reflected beam, so we have got to calculate the phase angle. In order to do this we have to be able to evaluate the structure factor, which in turn demands that we have at least approximate values of the atomic positions available. So we are still left with the problem of finding the atomic positions.

There are various pieces of information that can be used to give approximate atomic coordinates. The most useful is that, from past experience, we know roughly how far apart atoms in a molecule should be, and also how close two atoms in different molecules can approach. Bond lengths lie in the range 1 to 3 Å, whereas the closest approach of two atoms in different molecules, or the *van der Waals distance* is about twice this. The unit cell dimensions can often tell us something about the orientation of the molecules in the unit cell, especially if the molecule is planar. It is helpful if one particular X-ray reflexion is very much stronger than the rest, since there will be a correspondingly large amount of scattering matter in the planes responsible for this reflexion. Magnetic susceptibility measurements may help to define the orientation of a flat molecule as discussed in section 12.3. A determination of the refractive indices is sometimes of value, since the refractive index is small in a direction normal to the plane of a flat molecule and large along the length of a chain molecule.

In any particular crystal-structure determination a combination of these different factors must be taken into account, and evidence from as many sources as possible must be considered. Even so it is often still not possible to solve the crystal structure.

Use may then be made of *Patterson syntheses*, which have proved invaluable as an aid to crystal structure determinations. In this method a series similar to a Fourier series is summed, but the coefficients are the corrected intensities rather than the square roots of the corrected intensities

$$P(xyz) = \frac{1}{V} \sum_{-\infty}^{\infty} \sum_{-\infty}^{\infty} \sum_{-\infty}^{\infty} F_{hkl}^2 \cos 2\pi(hx/a + ky/b + lz/c).$$

This series, unlike equation (10), may be summed directly from the corrected intensities, since the coefficients are independent of the signs of the structure factors. A contour map can be plotted as before, and it can be shown that the peaks on the map correspond, not to the actual atomic positions, but to vector distances between pairs of atoms. Thus if a Patterson peak lies at the position x, y, z, there must be two atoms in the cell that are related both in distance apart and in direction in the same manner that the point x, y, z is related to the origin. If there are N atoms in the unit cell, there will be approximately N^2 Patterson peaks, and it is seldom possible to assign all these peaks to the correct pairs of atoms. Nevertheless, sufficient information may be gained to locate some of the atoms, particularly if there are one or two atoms in the molecule that are appreciably heavier than the rest. Once the positions of one or two atoms have been found with certainty, it becomes much easier to find the positions of the rest, either by interpretation of other Patterson peaks or by use of the methods described above.

Occasionally it is possible to solve a crystal structure directly. If the molecule contains one heavy atom, almost all the phases will be determined by this one atom alone. If, further, this heavy atom lies on a special position such that all three of its co-ordinates are predetermined, it is possible to calculate immediately the probable signs of all the structure factors. These signs can be used directly for a preliminary Fourier synthesis which will show the positions of the other atoms. The inclusion of these other atoms in the next set of calculated structure factors will show whether any of the phases are in fact opposite to those given by the heavy atom alone. It was in this way that one of the first complete structure determinations of a complex organic molecule, platinum phthalocyanine, was carried out.†

Sometimes it is possible to make use of a pair of isomorphous crystals. In one crystal an atom with a scattering factor f_1 will be located at a certain point. In the other crystal this atom will have been replaced by an atom with scattering factor f_2. Corresponding reflexions from the two crystals will differ in intensity, and these differences may be regarded as arising from a hypothetical crystal consisting only of atoms with a scattering power equal to $(f_1 - f_2)$. The structure of this hypothetical crystal can readily be determined, since there is only one atom to deal with. Once we have located this replaceable atom, we can calculate its contribution to each structure factor of the real crystals. By comparing the observed structure factors with the contributions from the replaceable atom, we can usually fix the sign of each structure factor. Once we know most of the signs, we can proceed as before.

Finally crystallographers have discovered that there are systematic relations between the phases of the stronger reflexions and that these relations may be used to obtain phases directly without any assumptions about the nature or the relative positions of the atoms in the unit cell. This so-called *direct method* of phase determination will undoubtedly assume an ever-increasing importance in the solution of crystal structures.

8.6. The temperature factor

The scattering factor curves shown in Fig. 1 were calculated on the assumption that the atoms were at rest. This can never be true under experimental conditions, partly because of the presence of zero-point vibrational motion (section 2.3) and partly because of the thermal vibration of the molecules at the temperature of the experiment. This motion of the atoms in the crystal has the effect of reducing the scattering factor still further as the Bragg angle increases. We may allow for this additional fall-off by multiplying the scattering factor by an exponential factor of the form

$$\exp\{-B(\sin\theta/\lambda)^2\},$$

† J. M. Robertson and I. Woodward, *J. chem. Soc.* 1940, 38.

where B is known as the *temperature factor*, and is related to the mean square amplitude of vibration of the atoms. Values of B are usually determined empirically from the fall-off in intensity of the X-ray reflexions at higher Bragg angles. Sometimes it is sufficient to use a single mean temperature factor for all the atoms in the unit cell, but for accurate work we must assign each atom its own individual temperature factor and, moreover, allow each atomic temperature factor to be different in different directions. The use of individual anisotropic temperature factors involves a substantial increase in the amount of computation, which has been dealt with only since the application of electronic computers to crystallographic problems.

8.7. The process of refinement

Suppose we have located approximate positions for all the atoms (except hydrogen atoms) by a combination of the methods discussed in section 8.5, and from these positions we have obtained a set of calculated structure factors by means of equations (1) and (3). If our proposed structure is correct, there will be rough agreement between the observed and calculated structure factors. A useful measure of this agreement is given by the *reliability* or *R factor* defined by

$$R = \frac{\sum |F_{hkl}^{\text{obs}} - F_{hkl}^{\text{calc}}|}{\sum |F_{hkl}^{\text{obs}}|}.$$

A trial structure will probably be correct if R has a value of about 0·35. Our next step is to transfer the signs from the calculated to the observed structure factors and to carry out a Fourier synthesis. From the atomic peaks on the contour map we can read off new atomic coordinates. We use these new atomic coordinates to calculate another set of structure factors, and check that our R factor has dropped, that is, that the agreement between the calculated and observed structure factors has got better. Some of the signs of the calculated structure factors may have changed, and we use this new set of signs to evaluate another Fourier synthesis which gives a better set of coordinates.

This cycling process is repeated until there are no more sign changes. At this point Fourier refinement must necessarily cease and R should have dropped to a value of about 0·15. The atomic coordinates obtained from the last cycle of Fourier refinement will not be very accurate because we have been using a finite number of terms in the series, because we have neglected any hydrogen atoms that the molecule may possess, and also because we have made no effort, at this stage, to refine the temperature factors.

In order to improve the agreement between observed and calculated structure factors after Fourier refinement has stopped, two main techniques are used. One method employs what is called a *difference synthesis*. This is similar to a normal Fourier synthesis except that, for the coefficients, we use the difference between corresponding pairs of observed and calculated structure factors. The resulting contour map will show us why there is not better agreement between the observed and calculated structure factors. Small alterations in the atomic coordinates may be suggested, anisotropic thermal motion of the atoms may be indicated, and the positions of the hydrogen atoms may be displayed. These new features can be incorporated in the next set of calculated structure factors and another difference map computed. This cycle of operations is repeated until there are no more characteristic features on the difference maps and until the R factor has been reduced as low as possible.

It should be noticed that difference syntheses automatically correct for most of the finite-series errors (section 8.4), since we may assume that the same errors that are present in a Fourier synthesis carried out with a limited number of observed structure factors will be present in a synthesis carried out with the same number of calculated structure factors. Thus in a synthesis calculated from the difference between the observed and calculated structure factors these finite-series errors will cancel.

Alternatively, a *least-squares analysis* may be carried out in which the sum of the squares of the differences between pairs of observed and calculated structure factors is minimized with respect to the structure parameters. The method is very powerful and successive cycles of least-squares refinement may be carried out on an electronic computer without any intervention by the operator. Moreover the observed structure factors can be assigned different weights depending on the reliability of their estimation, and, provided the observational errors are random, the least-squares method leads to a systematic and objective estimate of the errors in the different parameters.

Normally a combination of these two techniques is used: least-squares analysis to refine the positional and thermal parameters of all the heavier atoms; and a difference synthesis to locate hydrogen atoms and to show that there are no other significant regions of positive electron density. If visual intensities are employed, a final R factor of 0·06 to 0·12 is reasonable. These figures can be lowered to 0·02 to 0·04 with carefully obtained counter intensities.

8.8. The crystal structure of s-tetrazine†

We shall now illustrate many of the problems discussed in the last three chapters by considering in detail the determination of the crystal structure of s-tetrazine (Fig. 7.12). This is a relatively simple structure and was solved in projection. Copper radiation with a wavelength of 1·542 Å was used to take Weissenberg photographs of the $(h0l)$ and $(0kl)$ reflexions. The symmetry of the photographs showed that the cell was monoclinic, and measurement of the distance of several reflexions from the central line of the film gave the cell dimensions

$$a = 5·23 \pm 0·01, \quad b = 5·79 \pm 0·01, \quad c = 6·63 \pm 0·01 \text{ Å},$$
$$\beta = 115° 30' \pm 15'.$$

The special absences $(h0l)$ for l odd and $(0k0)$ for k odd gave the space group as $P2_1/c$. A reasonable density was calculated on the assumption of only two molecules in the unit cell. The presence of only two molecules in this space group demands that the molecule shall possess some symmetry, and the only possibility, as discussed in section 7.3, is that the molecule is centrosymmetric. The centre of each molecule must coincide

† F. Bertinotti, G. Giacomello, and A. M. Liquori, *Acta crystallogr.* 1956, **9**, 510.

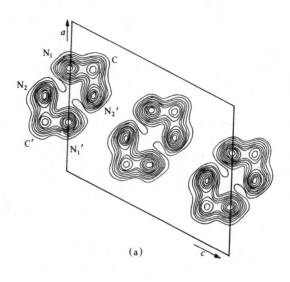

(a)

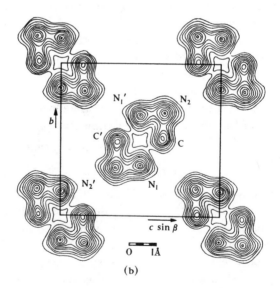

(b)

FIG. 4. Fourier maps of s-tetrazine: (a) electron density projected down [b], (b) electron density projected down [a]. Contours are drawn at intervals of 1 e.Å⁻², starting from 2 e.Å⁻².

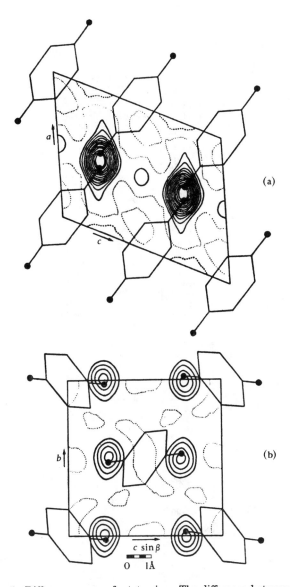

(a)

(b)

Fig. 5. Difference maps of s-tetrazine. The difference between the electron density obtained from the observed structure factors and that obtained from the calculated structure factors: (a) projected down $[b]$, (b) projected down $[a]$. The probable positions of the hydrogen atoms are indicated by black circles. Contour lines are drawn at intervals of 0.1 e.Å^{-2}, starting from $+0.3$ e.Å^{-2}.

TABLE XXXII

Observed and calculated structure factors of s-tetrazine

hkl	F_{hkl}^{obs}	F_{hkl}^{calc}	$\lvert F_{hkl}^{obs} - F_{hkl}^{calc} \rvert$
100	26·90	28·36	1·46
200	−14·96	−15·00	0·04
300	−11·42	−9·68	1·74
002	8·32	8·68	0·36
004	−5·26	−6·48	1·22
006	10·02	9·88	0·14
008	5·52	4·84	0·68
102	−14·20	−14·84	0·64
104	2·04	0·92	1·12
106	7·34	8·76	1·42
202	−13·46	−14·48	1·02
302	−6·44	−6·84	0·40
304	−2·32	−2·00	0·32
402	−4·58	−4·00	0·58
102	32·34	31·32	1·02
104	−11·42	−10·28	1·14
108	4·56	5·24	0·68
204	−8·00	−7·88	0·12
206	−6·88	−6·40	0·48
302	−9·62	−9·80	0·18
304	−11·48	−11·24	0·24
306	−4·94	−4·32	0·62
402	5·98	5·28	0·70
404	−5·90	−5·80	0·10
406	−2·80	−2·80	0·00
408	4·98	4·16	0·82
502	11·72	12·32	0·60
504	6·22	6·20	0·02
506	−3·96	−2·52	1·44
508	3·62	3·96	0·34
602	3·50	3·96	0·46
604	7·34	8·76	1·42
020	−5·52	−4·92	0·60
060	7·48	7·96	0·48
012	−15·12	−16·67	1·55
013	−7·14	−6·46	0·68
014	4·00	3·40	0·60
016	2·00	1·88	0·12
017	3·94	3·87	0·07
021	−12·82	−13·34	0·52
022	−7·96	−7·07	0·89
023	−3·06	−3·10	0·04
024	−4·44	−4·21	0·23
025	3·86	3·82	0·04
031	−11·50	−12·50	1·00
033	−5·20	−5·08	0·12
035	−4·08	−4·41	0·33
041	4·26	3·70	0·56
042	−3·58	−3·63	0·05
044	−4·12	−3·99	0·13
051	6·78	6·58	0·20
053	−3·20	−3·00	0·20

with a centre of symmetry in the unit cell, and the asymmetric unit is one carbon atom, one hydrogen atom, and two nitrogen atoms. The X-ray intensities were estimated by visual comparison with a standard strip, and fifty-two independent reflexions were observed.

The orientation of the molecule was obtained from a consideration of the strong X-ray reflexions. From this deduced orientation and from the expected molecular parameters, a trial set of coordinates was obtained. These coordinates were used to calculate a set of structure factors which gave satisfactory agreement with the observed structure factors, thus confirming the proposed model. Several cycles of Fourier refinement were then carried out for each projection until there were no more sign changes. The R factor at this stage was 0·11. The final Fourier maps of the two projections are shown in Fig. 4.

Difference maps for each projection were then computed. They showed the need for slight alterations in the positions of the carbon and nitrogen atoms, and the presence of anisotropic thermal motion. These additional refinements were incorporated and the next difference maps, which are reproduced in Fig. 5, showed quite clearly the approximate position of the hydrogen atom. Finally the hydrogen atom was included in the calculated structure factors, reducing the R factor to 0·07. Table XXXII lists the observed structure factors, the final calculated structure factors, and the difference between each corresponding pair. The final atomic coordinates are shown in Table XXXIII. Since X-rays do not locate hydrogen atoms accurately, the hydrogen atom is not included in this table. The atomic coordinates yield the molecular parameters shown in Table XXXIV. It will be seen that the two chemically equivalent C—N bond lengths differ by 0·022 Å. This is the sort of error that can be expected in a good two-dimensional analysis of an organic molecule with no heavy atoms. Heavy atoms will make the errors rather greater, whereas a three-dimensional analysis of s-tetrazine would probably reduce the errors in the molecular parameters by about half.

TABLE XXXIII

Fractional coordinates of the atoms in a crystal of s-tetrazine

Atom	x/a	y/b	z/c
C	0·2546	0·0153	0·1380
N_1	0·1834	−0·1669	−0·0015
N_2	0·0849	0·1795	0·1486

TABLE XXXIV

Molecular parameters of s-tetrazine found by X-ray diffraction

$r_0(CN_1)$	1·345 Å	$N_2' N_1 C$	116° 14′
$r_0(CN_2)$	1·323	$N_1' N_2 C$	116° 24′
$r_0(N_1 N_2')$	1·321	$N_1 CN_2$	127° 22′

It should be noticed that all X-ray diffraction does is to locate atoms. It does not tell us how these atoms should be joined together to form a molecule. Only experience and an accumulation of chemical knowledge allow us to do this.

8.9. The crystal structure of 4, 5-dimethyl phthalic thioanhydride $C_{10}H_8O_2S$†

As an example of a rather more complicated molecule whose structure was finally decided by an X-ray analysis of the crystal we may consider 4, 5-dimethyl phthalic thioanhydride. Two possible formulae have been suggested for this molecule (Fig. 6). The object of the X-ray analysis was not to obtain accurate molecular parameters but to decide which of these two possible molecular configurations was correct. Hydrogen atoms were, therefore, ignored, a single isotropic temperature factor was used, and the structure solved in projection. The relative intensities of 200 reflexions from the $(0kl)$ planes and 60 from $(h0l)$ were estimated visually from multiple-film Weissenberg photographs taken with copper radiation. The systematic absences showed that the space group was $P2_1/c$. The cell dimensions, obtained with an accuracy of ±1 per cent from rotation photographs, were

$$a = 4·08, \quad b = 13·6, \quad c = 16·7 \text{ Å}, \quad \beta = 92°.$$

† W. T. Eeles, *Acta crystallogr.* 1956, **9**, 365.

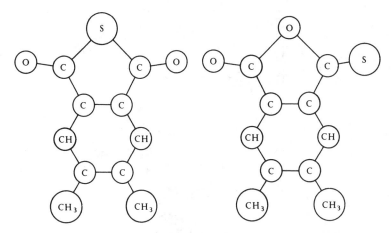

FIG. 6. Two possible structures for 4, 5-dimethyl phthalic thioanhydride.

The observed density indicated that there were four molecules in this unit cell: the asymmetric unit is therefore the whole molecule.

The reflexion from the (102) planes was by far the strongest, and it was presumed that the molecules lay roughly parallel to these planes. The (0kl) projection was solved by means of a two-dimensional Patterson synthesis. The resulting contour map is shown in Fig. 7. The sulphur atom has about twice the scattering power of any other atom in the molecule, and large peaks representing vectors between the sulphur atoms in adjacent molecules are to be expected. There should be three of these sulphur/sulphur peaks in the asymmetric unit of the Patterson projection, and those marked A, B, and C were finally selected. The peak A arises from sulphur atoms related by a centre of symmetry, and, from the coordinates of this peak, the y and z coordinates of the sulphur atom could be determined. With this is a start, other peaks could be explained and a trial structure obtained. It is very useful to have an atom about as heavy as sulphur in the molecule, since this one atom can be fairly readily located, and yet it is not so heavy as to obscure the scattering from the other atoms.

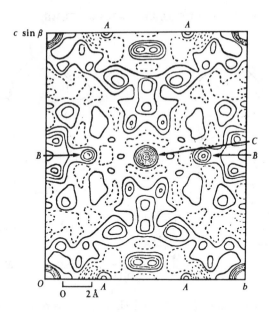

FIG. 7. (0*kl*) Patterson map of 4, 5-dimethyl phthalic thioanhydride. The scale is arbitrary and negative contours are broken. The peaks A, B, and C represent vectors between sulphur atoms in different molecules.

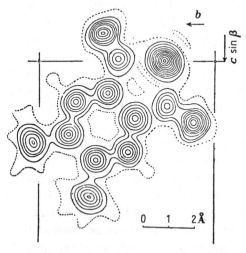

FIG. 8. (0*kl*) Fourier map of 4, 5-dimethyl phthalic thioanhydride. Contour lines are drawn at 1 e. Å^{-2} except at the sulphur atom where, above 4 e. Å^{-2}, they are at 2 e. Å^{-2}.

The trial structure was refined by successive Fourier and difference syntheses until the R factor had fallen to 0·17 for the 200 (0kl) reflexions. The final Fourier map of this projection is shown in Fig. 8. It indicates quite unambiguously that the molecule is not a thio-ketone but that the sulphur is incorporated in a five-membered ring.

Owing to the short a spacing, the atoms in the other two projections overlap rather badly and the Patterson maps would yield no useful information. Consequently, in order to solve the (0kl) projection, use was made of three pieces of information

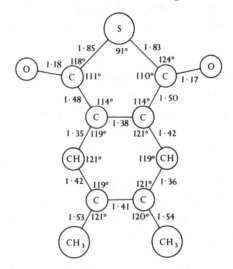

Fig. 9. Molecular parameters of 4, 5-dimethyl phthalic thioanhydride.

already available: the orientation of the molecule, the z co-ordinate of each atom, and standard values for the bond lengths. This information was sufficient to give a trial set of x coordinates. These coordinates were refined by a Fourier method until R dropped to 0·26. From the final x, y, and z coordinates, the molecular parameters shown in Fig. 9 were obtained.

This structure analysis is not as accurate as that of s-tetrazine discussed in the previous section, partly because of the presence of the relatively heavy sulphur atom, partly because two of the projections are badly resolved, and partly because fewer refinement cycles were carried out. Nevertheless, there is now no question of the arrangement of the atoms within the molecule, even though the internuclear distances, as shown by the internal consistency of the bond lengths in the benzene ring, are probably not accurate to more than ± 0.04 Å.

8.10. Limitations

X-ray diffraction can be used to study various other properties of solids apart from the atomic distribution. These topics are, however, beyond the scope of this book. In the realm of molecular-structure determination, there seems to be no immediate limit to the size of molecule that can be dealt with. The structure of vitamin B_{12}, a molecule with nearly 200 atoms, has been successfully solved, as have the structures of some proteins and enzymes, each with many hundreds of atoms.

X-ray analyses cannot yet compete with spectroscopic methods in accuracy, partly due to the inaccuracies in the observed intensities. A more serious drawback is the effect of molecular thermal motion which can lead to an apparent shortening of bond lengths amounting to several hundredths of an Ångstrom unit. Methods are being devised to allow for this effect, but only in very simple molecules of high symmetry can an adequate correction be made.

Another problem, especially in large molecules, is that it is advantageous, in order to solve the structure, to have a heavy atom present, but then the accuracy with which the positions of the other atoms can be determined is reduced. It is sometimes possible to get round this difficulty by use of two isomorphous substances, one of which contains a heavy atom. Once the structure of the crystal containing the heavy atom has been solved, the refinement can be carried out on the substance without the heavy atom. Potassium and ammonium salts, which are often isomorphous, are particularly useful for this purpose.

Hydrogen atoms can be located provided that the diffracted intensities have been measured with care. However hydrogen atoms become less easy to detect in the presence of heavy atoms since errors due to absorption, and uncertainty in the atomic scattering factors, become so great that hydrogen atoms are lost in the general background of the Fourier and difference maps. Moreover, even a carbon atom is a heavy atom compared with a hydrogen atom, and no accurate bond lengths involving hydrogen atoms have yet been found by X-ray diffraction.

The severest limitation in X-ray work used to be the amount of time and labour involved. The advent of automatic data-collection and data-processing has speeded operations up tremendously. It may still take a team of workers several years to determine the structure of a molecule containing hundreds or thousands of atoms, but frequently the difficulties are largely concerned with the preparation of suitable single crystals. The structure of a medium-sized molecule containing 20 to 50 atoms (excluding hydrogen atoms) can often be determined in a matter of weeks. But it must be remembered that the solution of a crystal structure is not automatic. There is no guarantee that the phase problem can be solved for any particular structure. If no solution can be found it is not possible to find the structure of the molecule.

REFERENCES

F. C. PHILLIPS, *An Introduction to Crystallography* (Longmans).

C. W. BUNN, *Chemical Crystallography* (Clarendon Press).

J. M. BIJVOET, N. H. KOLKMEIJER, and C. H. MACGILLAVRY, *X-Ray Analysis of Crystals* (Butterworth).

K. LONSDALE, *Crystals and X-Rays* (Bell).

R. W. JAMES, *X-Ray Crystallography* (Methuen).

H. LIPSON and W. COCHRAN, *The Determination of Crystal Structures* (Bell).

International Tables for X-Ray Crystallography (Kynoch Press).

IX

NEUTRON DIFFRACTION

9.1. Introduction

NEUTRON diffraction is potentially an extremely powerful method for molecular structure determination. In many ways the principles of the method are similar to those of X-ray diffraction, but there are some important differences which we shall consider in the next section. Neutron diffraction, like X-ray diffraction, can be used to study solids in the form of powders or single crystals and, again like X-ray diffraction, it is most effective when used with single crystals. Since sources of neutrons are scarce, neutron diffraction has been used mainly to investigate those aspects of molecular structure that cannot readily be ascertained by means of X-ray diffraction. Thus the results of the two methods are frequently complementary, and this is precisely why the results obtained by neutron diffraction have proved to be so interesting and important.

9.2. Differences between neutron and X-ray diffraction

Sources. One of the main limitations on the use of neutron diffraction is the need for an atomic pile to produce sufficient neutrons for diffraction purposes. The neutrons are released by atomic fission, which means that they have a very high initial velocity and, from the de Broglie equation (section 5.1), a very small initial wavelength. In order to increase the wavelength to about 1 Å, the value needed for diffraction by a crystal, we must slow down the neutrons by allowing them to collide with atoms in the pile. If the temperature of the pile is T, the neutrons will have a root mean square velocity v given by

$$\frac{M_n v^2}{2} = \frac{3kT}{2},$$

where M_n is the mass of the neutron. We may eliminate v from this equation and the de Broglie relation and obtain

$$\lambda^2 = h^2/3M_n kT.$$

If $T = 0°$ C, the wavelength turns out to be 1·55 Å and, if $T = 100°$ C, the wavelength is 1·33 Å. This is just the range needed for diffraction work, and fortunately, corresponds to the temperature convenient for the operation of an atomic pile. If, therefore, we allow a beam of neutrons to escape from a pile, they will emerge with a spectrum of wavelengths centred around 1·5 Å. In order to monochromatize this beam it is customary to reflect it from a single crystal, in which case the diffracted beam will have a wavelength given by

$$\lambda = 2d_{hkl}\sin\theta_{hkl},$$

where, as usual, d_{hkl} is the interplanar spacing and θ_{hkl} is the Bragg angle. Convenient single crystals to use are calcite or metallic lead. The (220) planes of calcite give a mono-chromatized beam with $\lambda = 1·16$ Å, whereas the (111) planes of lead give $\lambda = 1·08$ Å.

Although the diffracted beam is said to be monochromatic, this word is used rather loosely when applied to neutron beams, since there is nothing that corresponds to the characteristic radiation of X-rays (section 6.2). Consequently, the mono-chromator actually selects a small band of wavelengths from the incident beam, the width of the band being controlled mainly by the angle of divergence of the incident beam.

A beam of neutrons, even from a pile, is weak compared with a beam of X-rays from an ordinary X-ray tube, and, in order to get detectable diffraction effects, we have to work with a beam which is ten or twenty times as broad as an X-ray beam. The size of the specimen, whether single crystal or powder, must be correspondingly large. Attempts are being made to increase the number of neutrons available from a pile and to reduce the dimensions of the beam and specimen. There is little doubt that these objectives will be achieved and that the size factor will not finally be a severe limitation on neutron diffraction work.

Detectors. X-rays can be recorded photographically or they can be detected by various sorts of counters (section 8.2). Neutrons can be detected only by counters. Owing to the size of neutron beams and to the band of wavelengths selected by the monochromator, neutron counters are low in resolution, and the size of unit cell that can be investigated is limited. Moreover, neutron counters need to be adequately shielded against spurious radiation with the result that they are massive in construction and unwieldy in operation. All these problems are quite serious ones, but they will diminish as the size of the neutron beam is reduced.

Scattering factors. X-rays are scattered primarily by electrons. Except for magnetic materials, which we shall discuss in section 9.5, neutrons are scattered by nuclei. Atomic scattering factors for neutrons do not vary directly with atomic number as do scattering factors for X-rays (section 8.1), and all atoms scatter neutrons equally well within a factor of 2 or 3. Unlike X-ray scattering factors, neutron scattering factors cannot be calculated, and it is necessary to obtain each one empirically. There is no dependence of neutron scattering factors on θ, the Bragg angle. The reason for this last difference between neutrons and X-rays is that a nucleus is small compared with the wavelength of the neutrons, whereas a cloud of extranuclear electrons is of the same dimensions as the wavelength of X-rays. The scattering factors of the carbon atom for X-rays and neutrons are compared in Fig. 1. Table XXXV shows comparative scattering factors for a few different elements. Two important points can be seen from this table. In the first place the scattering factor for neutrons depends on the particular isotope. For most elements the scattering factor for each isotope has not been measured, and, except for hydrogen and carbon, Table XXXV lists the mean scattering factor for the naturally occurring isotopic mixture. Secondly, some of the neutron scattering factors are negative. When X-rays are scattered by atoms, there is always a phase change of 180° between the incident and scattered waves. The same is true when most nuclei scatter neutrons, but there are a few for which there is no phase change. Convention has decided that these few should be given negative scattering factors, and that the remaining neutron scattering factors and all X-ray scattering factors should be positive.

Since all neutron scattering factors have roughly the same

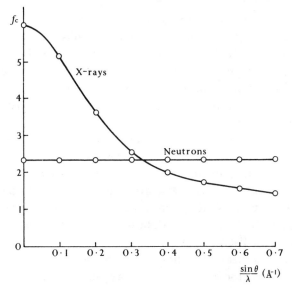

FIG. 1. Comparison of the neutron and X-ray scattering factor
curves of the carbon atom.

TABLE XXXV

*A comparison of the neutron and X-ray scattering factors
of certain atoms*†

Nucleus	Neutron scattering factor ⋅	X-ray scattering factor at $\theta = 0$
H^1	−1·35	1
H^2	2·31	1
C^{12}	2·35	6
C^{13}	2·14	6
O	2·11	8
Na	1·24	11
Cl	3·52	17
Mn	−1·32	25
Cu	2·70	28
I	1·85	53
Pt	3·38	78
Pb	3·41	82
U	3·02	92

† Adapted from G. E. Bacon, *Neutron Diffraction* (Clarendon Press).

values, there are not the same restrictions on the nature of compounds capable of investigation as there are with X-rays. It is easier to find a hydrogen atom in a substance containing uranium by means of neutron diffraction than it is to find a hydrogen atom in a substance containing carbon by means of X-ray diffraction. For this reason the main use of neutron diffraction in the determination of molecular structures has been for the location of hydrogen atoms.

The use of Fourier syntheses. Since neutron scattering factors do not fall off at higher Bragg angles as do X-ray scattering factors, the errors due to the use of a finite number of terms in a Fourier synthesis (section 8.4) are much greater with neutrons than with X-rays. With neutrons the series is neither infinite nor converging. This difficulty can be overcome by the use of difference maps as discussed in section 8.7.

Apart from these differences between the diffraction of X-rays and neutrons, the methods of X-ray diffraction and neutron diffraction are very similar, and we have already discussed the necessary principles in the chapters on X-ray diffraction. We can, therefore, immediately give some examples of the use of neutron diffraction in the elucidation of structural problems.

It must be remembered that, because of the scarcity of neutron sources, there are relatively few workers in the field of neutron diffraction compared with X-ray diffraction. It is necessary, therefore, to choose the type of problem that is worth investigating with great care. No one would use neutron diffraction to obtain, say, the structure of potassium nitrate, because it could be done much more readily and just as accurately by means of X-ray diffraction. Only if the structure has some aspects that cannot readily be settled by means of X-ray diffraction is it worth while going to the trouble of collecting neutron diffraction intensities. For this reason most neutron diffraction studies have been concerned with the location of hydrogen atoms.

9.3. The structure of sodium sesquicarbonate†

Sodium sesquicarbonate, $Na_2CO_3.NaHCO_3.2H_2O$, is a

† G. E. Bacon and N. A. Curry, *Acta crystallogr.* 1956, **9**, 82.

particularly suitable substance for investigation by neutron diffraction. Although an earlier X-ray examination† had located all the heavier atoms and had shown the existence of a short hydrogen bond joining two carbonate groups, the details of this bond and also the disposition of the hydrogen atoms in the water molecule were not clear. Moreover, the cell dimensions showed that the desired information should be obtainable from a two-dimensional projection down [b], which is the shortest axis. Thus only the intensities of the (h0l) reflexions need be measured. This is a great help because it is difficult to collect three-dimensional neutron intensities from single crystals with large unit cells. Finally, the geometry of a neutron diffraction spectrometer requires the crystal to be elongated in the vertical direction, which is [b] in this case. Large single crystals of sodium sesquicarbonate can be grown with [b] as the direction of elongation.

The X-ray examination had shown that there were four formula units in a monoclinic cell with dimensions

$$a = 20\cdot41, \quad b = 3\cdot49, \quad c = 10\cdot31 \text{ Å}, \quad \beta = 106\cdot3°.$$

The general absences of the (hkl) reflexions with (h+k) odd showed that the lattice was centred on the C face. The absence of (h0l) reflexions with l odd showed the presence of a c glide perpendicular to [b]. These absences are consistent with two possible space groups, Cc and C2/c. The latter is centrosymmetric with eight general positions, and was confirmed as the true space group, partly by the external shape of the crystals but mainly by a successful structure determination. The space group C2/c requires one of the three sodium ions and the hydrogen atom of the bicarbonate group to lie in special positions. The complete structure determination showed that the odd sodium ion lies on a twofold axis. The hydrogen atom must therefore lie on a centre of symmetry. All this information was known before the start of the neutron diffraction investigation.

The intensities of 225 different (h0l) reflexions were measured with a neutron wavelength of 0·81 Å. The size of the crystal was 3×4×11 mm, the last being in the direction of [b], which

† C. J. Brown, H. S. Peiser, and A. Turner-Jones, *Acta crystallogr.* 1949, **2**, 167.

was set vertical. From the known positions of the sodium, carbon, and oxygen atoms, shown in Fig. 2, structure factors were calculated. The signs of these structure factors were transferred to the observed structure factors and two one-dimensional difference syntheses (sections 8.4 and 8.7) were carried out with the $(h00)$ and $(00l)$ reflexions. These simple difference maps were

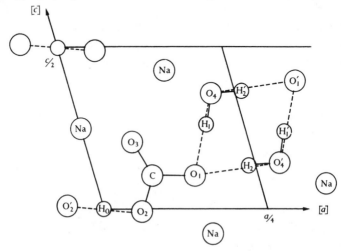

FIG. 2. The labelling of the atoms in sodium sesquicarbonate.

sufficient to indicate approximate positions of the hydrogen atoms. The latter were included in the next set of calculated structure factors, and, with the resulting signs, a two-dimensional Fourier synthesis was computed. The Fourier map is shown in Fig. 3. Sodium, carbon, and oxygen atoms appear as peaks of about equal weight, but hydrogen atoms, because they have negative scattering factors for neutrons, appear as troughs, roughly as deep as the peaks are high. Different isotropic temperature factors were then included for the different atoms and a final difference synthesis, from which all but the hydrogen atoms had been subtracted, computed. The final difference map is shown in Fig. 4 in which the hydrogen atoms stand out very clearly. The final R factor was $0 \cdot 11$. The internuclear distances which involve hydrogen atoms are given in Table XXXVI, where they are compared, when possible, with the X-ray results.

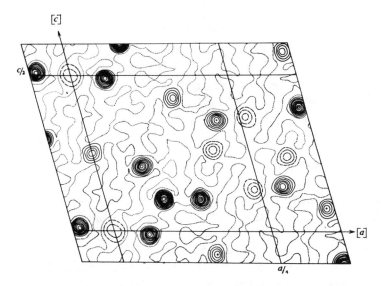

FIG. 3. (h0l) Fourier map of sodium sesquicarbonate. Projection of the neutron scattering density down [b]. Contours are drawn at equal arbitrary intervals with negative contours broken and the zero contour dotted.

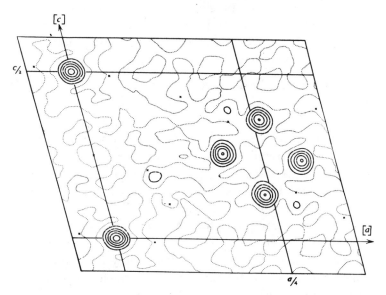

FIG. 4. (h0l) difference map of sodium sesquicarbonate. The heavy atoms, whose positions are indicated by crosses, have been subtracted, leaving only the hydrogen atoms. Contours are drawn at half the interval used in Fig. 3.

TABLE XXXVI

Internuclear distances in sodium sesquicarbonate

Bond	Neutron diffraction	X-ray diffraction
$O_1 \ldots O_4$	$2 \cdot 76 \pm 0 \cdot 02$ Å	$2 \cdot 72 \pm 0 \cdot 03$ Å
H_1——O_4	$1 \cdot 03 \pm 0 \cdot 03$	..
$O_1 \ldots O_4'$	$2 \cdot 78 \pm 0 \cdot 02$	$2 \cdot 77 \pm 0 \cdot 03$
H_2——O_4'	$0 \cdot 99 \pm 0 \cdot 03$	..
$O_2 \ldots O_2'$	$2 \cdot 50 \pm 0 \cdot 02$	$2 \cdot 53 \pm 0 \cdot 02$

Several conclusions may be drawn from the neutron diffraction results. Firstly, the neutron-diffraction investigation confirms the results of the previous X-ray investigation. Secondly, the hydrogen atoms of the water molecules are displaced slightly from the lines of the O ... O hydrogen bonds in such a way as to make the HOH angle approximately tetrahedral. Thirdly, these two hydrogen bonds are typical *long bonds* with the hydrogen atoms much nearer one oxygen atom than the other. Fourthly, it is not possible to decide whether the anisotropy of the contours of the hydrogen atom H_0 on the centre of symmetry is due to an atom lying in a single broad symmetric potential minimum, or whether the anisotropy is due to a disordered arrangement of hydrogen atoms each placed $1 \cdot 12$ Å from one or the other oxygen atom.

9.4. The structure of deuteroammonia†

Deuteroammonia, prepared from Mg_3N_2 and $99 \cdot 5$ per cent D_2O, does not form single crystals of a size suitable for neutron diffraction, but a very accurate investigation has been made of a powder sample weighing $3 \cdot 1$ g maintained at liquid nitrogen temperature $(-196°$ C$)$ in a thin-walled aluminium cup. Deuteroammonia is cubic with $a = 5 \cdot 073$ Å. The space group is $P2_13$ with four molecules in the unit cell. The molecule is required by the space group to have a threefold axis of symmetry, and the asymmetric unit is, therefore, one hydrogen atom and one-third of a nitrogen atom. Only four positional parameters need to be determined, three for the hydrogen atom and one for the nitrogen atom.

† J. W. Reed and P. M. Harris, *J. chem. Phys.* 1961, **35**, 1730.

Intensities of reflexions up to a maximum value of

$$h^2 + k^2 + l^2 = 22$$

were measured very carefully with a rotating sample and neutrons of wavelength $1 \cdot 113 \pm 0 \cdot 005$ Å. The structure was solved by trial and error, and refined by least-squares analysis with individual isotropic temperature factors. The agreement between the corrected and calculated intensities is shown in Table XXXVII. The positional parameters agree well with those obtained in a parallel study by X-ray diffraction of ordinary ammonia.[†] The two sets of coordinates are compared in Table XXXVIII. As would be expected, the position of the nitrogen atom is determined with less accuracy by neutron diffraction than by X-ray diffraction, but the position of the hydrogen atom is found with considerably greater accuracy by neutron diffraction.

TABLE XXXVII

Comparison of corrected and calculated intensities of deuteroammonia

$h^2 + k^2 + l^2$	Icorr	Icalc
2	11·86	12·29
3	160·63	162·15
4	76·28	72·52
5	11·19	12·04
6	18·99	18·50
8	28·87	31·03
9	137·60	132·78
10	84·80	87·03
11	19·53	18·38
12	23·00	26·23
13	16·79	16·88
14	52·20	53·50
16	0·00	0·06
17	22·59	20·34
18	102·33	106·86
19	21·09	17·87
20	0·00	1·75
21	183·60	178·40
22	49·95	60·85

[†] J. Olovsson and D. H. Templeton, *Acta crystallogr.* 1959, **12**, 832.

TABLE XXXVIII

*Comparison of positional parameters obtained by neutron
diffraction with those obtained by X-ray diffraction*

Atom	Parameter	Neutron diffraction	X-ray diffraction
N	$x = y = z$	$0\cdot2127 \pm 0\cdot0021$	$0\cdot2099 \pm 0\cdot0006$
H	x	$0\cdot3740 \pm 0\cdot0019$	$0\cdot40$
H	y	$0\cdot2632 \pm 0\cdot0037$	$0\cdot26$
H	z	$0\cdot1094 \pm 0\cdot0016$	$0\cdot11$

The neutron diffraction parameters yield a value of
$1\cdot005 \pm 0\cdot023$ Å for the length of the N—D bond, and a value of
$110\cdot4 \pm 2\cdot0°$ for the DND angle. These values may be compared

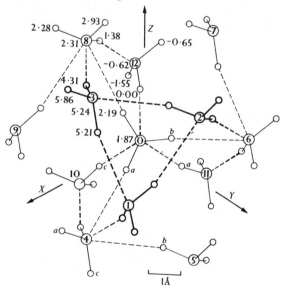

FIG. 5. The structure of ammonia viewed down [111] of the
cube. A central molecule (No. 0) is shown with its six
nearest neighbours (Nos. 4, 6, 8, 10, 11, 12) and its six
next-nearest neighbours (Nos. 1, 2, 3, 5, 7, 9). Nos. 0, 4, 5,
6, 7, 8, 9 lie roughly in a plane, whereas 1, 2, and 3 lie above
and 10, 11, and 12 lie below the plane. Hydrogen bonds are
indicated by dashed lines. Distances of some atoms from
the plane $x = y = z = 0$ are given in Å.

with $1 \cdot 008 \pm 0 \cdot 004$ Å and $107 \cdot 4 \pm 0 \cdot 2°$ obtained for free ammonia molecules by microwave spectroscopy.[†] The bond lengths agree within experimental error, but the bond angle is significantly greater in the solid than it is in the free molecule. Hydrogen bonds of length $3 \cdot 352 \pm 0 \cdot 011$ Å between nitrogen atoms form an interlocking network in the solid state as shown in Fig. 5.

9.5. Magnetic scattering

Although the potentialities of neutron diffraction in the field of molecular structure determination are enormous, the most important contributions to our knowledge so far have been connected with the structure of magnetic materials.

Apart from the scattering of neutrons by atomic nuclei, there is an additional scattering if the specimen contains paramagnetic atoms or ions. This magnetic scattering is due to the interaction of the magnetic moments of the neutrons with the permanent magnetic moments of the paramagnetic atoms. An atom has a permanent magnetic moment when one of its inner electron shells is only partially filled (section 12.4). Since the inner shells have a size comparable with the wavelength of the neutrons, the magnetic scattering factor will, unlike the nuclear scattering factor, fall off with increasing Bragg angle in the same manner that X-ray scattering factors do (section 8.1).

In general the magnetic moments of atoms in a paramagnetic crystal are arranged randomly. The magnetic scattering is, therefore, incoherent and forms a diffuse background which falls off rapidly at higher Bragg angles. However, there are two main classes of materials in which the individual magnetic moments are arranged in a regular way. In *ferromagnetic* substances the moments of the atoms are arranged parallel, whereas in *antiferromagnetic* substances the atoms are arranged in such a way that the magnetic moments of nearest neighbours are opposed. Ferromagnetic materials are usually transition metals such as iron, cobalt, or chromium, and antiferromagnetic materials are usually ionic salts of transition metals such as ferric oxide, Fe_2O_3, or manganous fluoride, MnF_2.

[†] M. T. Weiss and M. W. P. Strandberg, *Phys. Rev.* 1951, **83**, 567.

Neutron diffraction studies can discriminate between these different magnetic structures. For instance, manganous oxide. MnO, has been shown by X-ray diffraction to have the same crystal structure as NaCl (Fig. 6.21). Both the manganous and the oxygen ions lie in a face-centred array with the two arrays displaced in the same manner as in NaCl (section 8.3). The length of the unit cell side is 4·426 Å. In a neutron diffraction study below 120° K extra reflexions were found which indicated

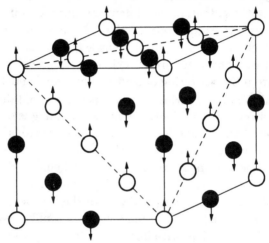

FIG. 6. The magnetic unit cell of manganous oxide, MnO. Only the manganous ions are shown. Parallel to the (111) planes the manganous ions form sheets in which all the magnetic moments are aligned in the same direction.

that the unit cell side should be twice this value.† These additional reflexions disappeared at temperatures above 120° K. The intensity of these *magnetic lines* could be satisfactorily explained if it was assumed that below 120° K the nearest neighbour manganous ions had their magnetic moments antiparallel as shown in Fig. 6, in which only the manganous ions are included.

REFERENCE

G. E. Bacon, *Neutron Diffraction* (Clarendon Press).

† C. G. Shull, W. A. Strauser, E. O. Wollan, *Phys. Rev.* 1951, **83**, 333.

PART III

Miscellaneous Methods

X

CLASSICAL STEREOCHEMICAL METHODS

10.1 Introduction

MANY simple molecules and ions are of the form AX_n, in which the central atom A is surrounded by n atoms of the type X. X can equally well be a group of atoms such as CH_3, CN, or NO_2, but for convenience we shall refer to X as though it were a single atom. We shall also, in this chapter, use the word molecule to describe the entity AX_n, even though it possesses a charge. For a long time chemists have thought that the disposition of the X atoms about the central atom would be symmetrical with the X atoms as far apart as possible. To a certain extent this view has been confirmed, and it has been almost entirely confirmed if we remember that a lone pair of electrons can sometimes take the place of one of the peripheral atoms. We met an example of this property of lone pairs of electrons when we discussed the structure of SF_4 in section 4.7.

Nowadays it is customary to relate the shape of an AX_n molecule to the atomic orbitals that the atom A are presumed to use in the formation of the bonds A—X. Table XXXIX lists some of the possible shapes in terms of these orbitals and of the coordination number n. Whenever orbitals of two or more different symmetries are employed, as in d^2sp^3, we say that *hybridization* occurs to produce a corresponding number of equivalent orbitals. Thus if the atomic orbitals are sp^3, four equivalent sp^3-hybridized orbitals pointing towards the corners of a regular tetrahedron are formed, and if they are d^2sp^3, six equivalent d^2sp^3-hybridized orbitals pointing towards the corners of a regular octahedron result. We are not concerned with the theoretical justification for the concept of hybridization, and have introduced it merely because it provides a convenient language with which to describe the molecules.†

TABLE XXXIX

Atomic orbitals and molecular shapes

Coordination number n	Most common orbitals of A used in bonding	Shape of AX_n molecule	Example
2	sp p^2	Linear Bent	CO_2 H_2O
3	sp^2 p^3	Plane trigonal Pyramidal	BCl_3 NH_3
4	sp^3 dsp^2	Tetrahedral Square planar	CH_4 $[PtCl_4]^{2-}$
5	dsp^3	Trigonal bipyramidal	PF_5
6	d^2sp^3	Octahedral	SF_6
8	d^4sp^3	Dodecahedral	$[Mo(CN)_8]^{4-}$

Chemical experiments had been used to obtain information about the shape of four- and six-coordinated molecules before the more modern physico-chemical methods had been devised or developed. The chemical arguments are based on the evidence of either geometrical isomerism or stereoisomerism. We shall consider these two sorts of isomerism in turn.

10.2. Geometrical isomerism

The two most likely shapes for a molecule AX_4 are square planar or tetrahedral. If the structure were planar, we would expect to get two geometrical isomers of a molecule such as AX_2Y_2, a *trans* form (Fig. 1 (a)) and a *cis* form (Fig. 1 (b)). These two forms are clearly different since the *trans* form has mmm (D_{2h}) symmetry whereas the *cis* form has only two mirror planes and has the lower mm (C_{2v}) symmetry. On the other hand, a tetrahedral structure would result in only one isomer for an

† For a discussion of hybridization see C. A. Coulson, *Valence* (Clarendon Press), Chapter 8.

AX_2Y_2 molecule since, however we interchange the X and Y atoms, we end up with a molecule of mm (C_{2v}) symmetry, and the various possibilities differ only by a reorientation of the whole

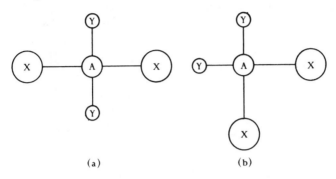

(a)   (b)

FIG. 1. The two forms of a square-planar molecule AX_2Y_2: (a) the *trans* isomer of mmm (D_{2h}) symmetry, (b) the *cis* isomer of mm (C_{2v}) symmetry.

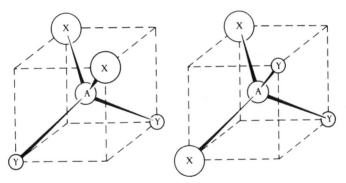

FIG. 2. The only form of a tetrahedral molecule AX_2Y_2. The two molecules shown differ only in orientation.

molecule (Fig. 2). Platinum complexes of the type PtX_2Y_2 often give two isomers, which suggests that the valencies of the platinum atom are distributed square planar. On the other hand, two isomers of a substituted methane CX_2Y_2 have never been isolated, which suggests that the bonding orbitals of a carbon atom attached to four univalent atoms are always disposed tetrahedrally. However, neither for the platinum nor for the carbon compounds can the chemical evidence be considered conclusive. The isolation of two isomers would be conclusive evidence for a square-planar structure if we could be sure that the isomerism is actually geometric and is not due to some other cause. It is always difficult, without appealing to physicochemical methods, to be sure that this is the case. For instance, the reaction between methyl iodide and tellurium yields the red α-dimethyl tellurium diiodide, α-$(CH_3)_2TeI_2$. If the latter is treated with silver oxide, a dihydroxy compound is formed which on dehydration and treatment with hydriodic acid gives a green compound.

$$Te + 2CH_3I \xrightarrow{\hspace{2cm}} \alpha\text{-}(CH_3)_2TeI_2 \text{ (red)}$$

$$Ag_2O$$

$$(CH_3)_2Te(OH)_2 \xrightarrow{(-H_2O)} (CH_3)_2TeO \xrightarrow{HI} \beta\text{-}(CH_3)_2TeI_2 \text{ (green)}$$

The green product was originally classed as β-$(CH_3)_2TeI_2$, and the α- and β-compounds were presumed to be *trans* and *cis* isomers.† The existence of these two isomers would imply a square-planar distribution of the valencies of tellurium. However, later work‡ has shown that the β-isomer is not monomeric, but is a salt $[(CH_3)_3Te][CH_3TeI_4]$, so we are left with only one geometrical isomer again. The existence of only one geometrical isomer suggests a tetrahedral distribution of valencies, but again the evidence is not conclusive, because the compound might really be square planar with one of the geometrical isomers too unstable to be isolated.

The same sorts of difficulty are met in molecules of the type AX_6. The three most probable shapes are plane hexagonal, trigonal prismatic, and octahedral, shown in Fig. 3. The number of possible geometrical isomers of substituted AX_6 molecules is given in Table XL. Of the many compounds of the type AX_4Y_2 and AX_3Y_3 that have been investigated none has given more than two isomers, which suggests very strongly that an octahedral configuration for AX_6 molecules is very common if not universal. Again, however, the failure to isolate a

† R. H. Vernon, *J. chem. Soc.* 1920, 86, 889; 1921, 687.
‡ H. D. K. Drew, ibid. 1929, 560.

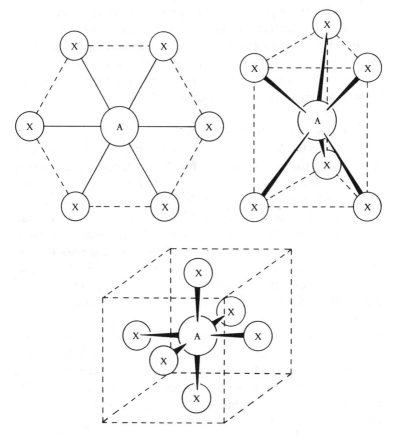

Fig. 3. The three most probable shapes of a molecule AX_6.

TABLE XL

Geometrical isomers of substituted AX_6 molecules

Compound	Plane hexagonal	Trigonal prismatic	Octahedral
AX_5Y	1	1	1
AX_4Y_2	3	3	2
AX_3Y_3	3	3	2

third isomer does not necessarily mean that the isomer does not exist, but merely that it may be very unstable. We must conclude, therefore, that counting the number of geometrical isomers of any one particular compound is not likely to be a very rewarding occupation. It is of no value for molecules with a tetrahedral or octahedral configuration, because such molecules have fewer geometrical isomers than the other, less regular possibilities. And it is of little value for the less regular shapes, because of uncertainty in the nature of the isomerism.

10.3 The use of bidentate groups

Although for any one particular compound the method described in the last section is of little use, the accumulation of consistent results for AX_6 molecules left little doubt that most AX_6 molecules were octahedral. The next question to be settled was which of the two isomers of AX_4Y_2 or AX_3Y_3 molecules was the *cis* and which the *trans* form. In order to answer this question, use was made of *bidentate groups*. These are molecules or ions that can occupy two of the coordination positions of the central atom. Examples of such bidentate groups are ethylenediamine, $NH_2.CH_2.CH_2.NH_2$, in which each nitrogen atom can occupy a coordination position, and the oxalato group, $O.CO.OC.O$, in which one oxygen atom from each half of the ion can co-ordinate to the central atom. The use of these bifunctional groups depends first of all on the assumption that the groups can occupy only adjacent coordination positions and thus necessarily form a *cis* compound. This must be the case with many of the commoner bidentate groups since they are not large enough to span the *trans* positions.

As an example of the use of bidentate groups we may consider the determination of the structure of derivatives of carbonato-tetramminocobaltic nitrate, $[Co(NH_3)_4(CO_3)](NO_3)$, whose cation is shown in Fig. 4. This salt forms no precipitate with barium chloride so that the carbonato group must be firmly attached to the cobalt atom. Moreover, the size of the carbonato group definitely excludes the possibility that the ion has the *trans* configuration. If this salt is treated with hydrochloric acid, carbon dioxide is evolved and two chloride ions replace the carbonato group. It is clear that the resulting dichlorotetramminocobaltic complex ought to be the *cis* compound (Fig. 5).

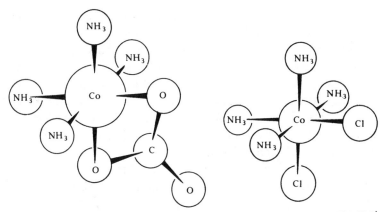

FIG. 4. The ion $[Co(NH_3)_4(CO_3)]^+$.

FIG. 5. The ion $[Co(NH_3)_4Cl_2]^+$.

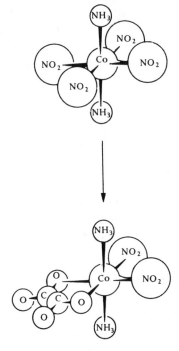

FIG. 6. The sole isomer of $[Co(NH_3)_2(NO_2)_2(C_2O_4)]^-$ obtained from the *trans* form of $[Co(NH_3)_2(NO_2)_4]^-$.

Unfortunately this conclusion is not justified, since it has been shown that considerable amounts of the *trans*-dichloro compound are obtained. Molecular rearrangements of this sort seem to occur quite frequently, and they obviously invalidate any conclusions about the stereochemistry of the product. Chemists are gradually learning how to minimize such rearrangements and, in the present example, little *trans*-dichloro compound is formed if the solid carbonato complex is allowed to react with an alcoholic solution of hydrogen chloride.

The same sort of argument has been applied in reverse to determine the structure of ammonium diamminotetranitro-cobaltate, $NH_4[Co(NH_3)_2(NO_2)_4]$.† Only one form of this salt has been isolated and an attempt was made to find the configuration chemically by treatment with oxalic acid to give $NH_4[Co(NH_3)_2(NO_2)_2(C_2O_4)]$. If the original salt were of the *trans* form, only one oxalato derivative could be formed (Fig. 6), whereas a *cis* compound would yield two derivatives (Fig. 7). Two derivatives were in fact isolated, but the obvious conclusion that the original salt was of the *cis* form has been put in doubt by an X-ray study‡ of the corresponding silver salt, $Ag[Co(NH_3)_2(NO_2)_4]$. The space group requires the complex ion to have 42 (D_4) symmetry so that the ion must have a fourfold axis. The two ammine groups must lie on this axis and the configuration is, therefore, *trans*. Here again, unless the ion has a different configuration in the solid state, molecular rearrangement must have taken place. There are many examples in which this use of bidentate groups has subsequently been shown to have given the correct structure, but without some additional evidence the possibility of rearrangement always casts doubt upon the conclusions.

10.4. Stereoisomerism

The uncertainties that surround the use of geometrical isomerism are largely removed by the use of stereoisomerism. For example, the fact that compounds of the form PtX_2Y_2

† W. Thomas, *J. chem. Soc.* 1922, 2069; 1923, 617.

‡ A. F. Wells, *Z. Kristallogr.* 1936, **95**, 74.

often occur in two forms strongly suggests that the valencies of the platinum atom are square planar. This was proved by a most ingenious and elegant chemical experiment which involved the preparation of *meso*-stilbenediamino*iso*butylenediaminoplatinous chloride.† If the valencies of the platinum atom are

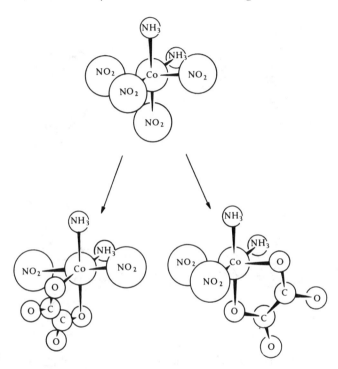

FIG. 7. The two isomers of $[Co(NH_3)_2(NO_2)_2(C_2O_4)]^-$ obtained from the *cis* form of $[Co(NH_3)_2(NO_2)_4]^-$.

arranged tetrahedrally (Fig. 8 (*a*)), the ion will possess a plane of symmetry which lies in the plane of the paper in Fig. 8 (*a*). The compound will therefore be optically inactive. On the other hand, if the valencies are arranged square planar (Fig. 8 (*b*)), the cation will possess no symmetry elements and should be optically active. In fact the compound could be resolved into enantiomorphous optically active forms by means of diacetyltartaric

† W. H. Mills and T. H. H. Quibell, *J. chem. Soc.* 1935, 839.

acid, thus proving the square-planar disposition of the four platinum valencies. It would, of course, have been equally feasible to prepare a similar compound in which the square-planar form had a plane of symmetry and the tetrahedral form no symmetry. We would then observe no optical activity but

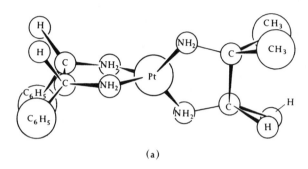

(a)

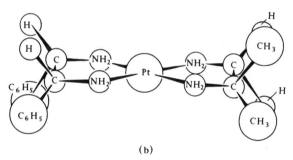

(b)

FIG. 8. The effect of the distribution of the valencies of platinum on the symmetry of *meso*-stilbenediamino*iso*butylenediamino-platinous chloride: (a) the tetrahedral distribution gives a molecule with a mirror plane, (b) the square-planar distribution gives a molecule with no symmetry.

this would prove nothing, because we would not know whether the lack of activity was due to the fact that the molecule possessed symmetry or whether the two optically active forms of the tetrahedral ion were too rapidly interconvertible. In other words, an asymmetric molecule does not necessarily show optical activity, and only positive answers are useful in experiments involving stereoisomerism.

Stereoisomerism can also be used to obtain information about the configuration of molecules containing an atom that is six-coordinated. An octahedral complex of the type [CoenX$_4$], where en stands for ethylenediamine, can exist in only one form (Fig. 9). If two bidentate groups are present, as in [Coen$_2$X$_2$], the *trans* form will possess a mirror plane (Fig. 10) and will be

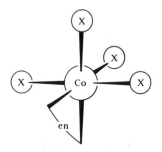

FIG. 9. The sole isomer of the complex [CoenX$_4$].

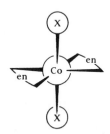

FIG. 10. The sole stereoisomer of *trans*-[Coen$_2$X$_2$].

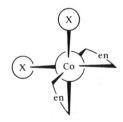
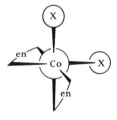

FIG. 11. The two stereoisomers of *cis*-[Coen$_2$X$_2$].

optically inactive. On the other hand the *cis* form does possess symmetry, but the only symmetry element is a twofold axis. Consequently, as we saw in section 1.3, enantiomorphous forms exist (Fig. 11) and optical activity is possible. Finally, if there are three bidentate groups, as in [Coen$_3$]$^{3+}$ the situation is the same as for the *cis* form of [Coen$_2$X$_2$] and enantiomers exist (Fig. 12). The isolation of the correct number of isomers with the requisite optical activity established conclusively the octahedral configuration of many six-coordinated atoms.

This procedure can, of course, be used to distinguish between *cis* and *trans* isomers of six-coordinated compounds with two bidentate groups. The method fails when only one isomer has been isolated and this cannot be resolved into optical enantiomorphs, since failure to resolve the compound, as mentioned before, does not necessarily mean that the stereoisomers do not exist.

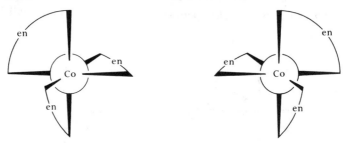

FIG. 12. The two stereoisomers of $[Coen_3]^{3+}$.

If we return to the example of $NH_4[Co(NH_3)_2(NO_2)_4]$ given in the last section, there is little doubt that the final products were correctly diagnosed, since one of the two oxalato derivatives was optically active whereas the other was not. It can be seen from Fig. 7 that the oxalato derivative in which the two nitro groups are *trans* possesses a mirror plane and will be inactive. The other derivative has no symmetry and will exist in enantiomorphous forms. The argument broke down when it was assumed that it must necessarily be a *cis* compound that yielded these two products. Each step in a chemical argument of this sort must be examined very closely in order to see what assumptions are involved and whether these assumptions are justified.

10.5. The stereochemistry of the vinylidineamines

As a final illustration of the power and limitations of classical stereochemical arguments, we shall consider the stereochemistry of some analogues of allene, $CH_2:C:CH_2$. Theoretical considerations and investigations of the infra-red and Raman spectra of this hydrocarbon prove that the planes of the two CH_2 groups are at right angles and that the molecule has $\bar{4}2m$ (D_{2d}) symmetry (Fig. 13).† The question now arises: What happens to the shape

† K. N. Rao, A. H. Nielsen, and W. H. Fletcher, *J. chem. Phys.* 1957, **26**, 1572, and references given in this paper.

of the molecule if one CH_2 group is replaced by an NH group to give vinylidineamine, $CH_2:C:NH$? If the disposition of the double bonds is similar in the two molecules, and if the two bonds to the nitrogen atom are not collinear, we would expect to find vinylidineamine with the shape shown in Fig. 14. This molecule

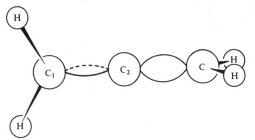

FIG. 13. The molecule of allene, $CH_2:C:CH_2$.

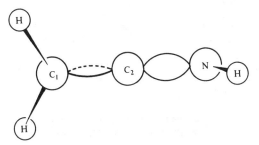

FIG. 14. The probable shape of vinylidineamine, $CH_2:C:NH$.

possesses a plane of symmetry and will be optically inactive. On the other hand a substituted vinylidineamine, $XYC:C:NZ$, where X and Y must be different and Z can be any atom, will possess no symmetry (Fig. 15) and should exist in enantiomorphous forms. However, if the valencies to the nitrogen atom are collinear, even this substituted vinylidineamine will possess a plane of symmetry (Fig. 16) and thus will be optically inactive. Attempts to solve this stereochemical problem failed for some time because simple vinylidineamines are too unstable to exist. In order to get round this difficulty it was decided to use a familiar stereochemical trick, and to replace the double bond $C_1{=}C_2$ by a saturated cyclohexane ring, which is stereochemically

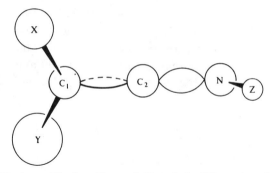

Fɪɢ. 15. A possible shape for a substituted vinylidineamine, showing the absence of symmetry.

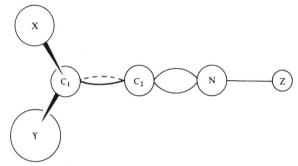

Fɪɢ. 16. A possible shape for a substituted vinylidineamine, showing the presence of a mirror plane.

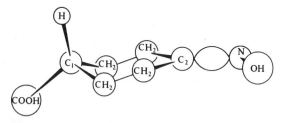

Fɪɢ. 17. The molecule of cyclohexanone-4-carboxylic acid oxime.

equivalent. Finally the oxime of cyclohexanone-4-carboxylic acid (Fig. 17) was prepared and shown to be resolvable into enantiomorphs.† A comparison of Fig. 15 and Fig. 17 will show that we are still dealing with the same stereochemical problem.

† W. H. Mills and A. M. Bain, *J. chem. Soc.* 1910, 1866.

So far the experimental evidence is that the oxime of cyclo-hexanone-4-carboxylic acid is optically active and therefore the bonds to the nitrogen atom cannot be collinear. This is incontrovertible. However, it has long been assumed that, as a consequence, a vinylidineamine would have a corresponding structure. This step involves the assumption that the double

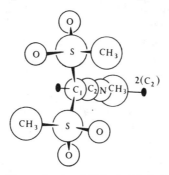

Fig. 18. The molecule of N-methyl-2:2-dimethyl-sulphonylvinylidineamine, showing the presence of a twofold axis.

bond $C_1{=}C_2$ has no effect on the adjacent bonds at the nitrogen atom, and it is now known that this assumption is not valid. The compound N-methyl-2:2-dimethylsulphonylvinylidine-amine, $(CH_3SO_2)_2C{=}C{=}N{-}CH_3$, has been prepared and the molecular structure determined by X-ray diffraction.[†] The space group is *Pbcn* and there are only four molecules in the unit cell. Since there are eight equivalent positions in this space group, the molecule must possess symmetry (section 7.3), and the alternatives are a centre of symmetry and a twofold axis. Since a centre of symmetry is quite impossible, the molecule must possess a twofold axis, and the bonds to the nitrogen atom must, therefore, be collinear (Fig. 18). All we have now proved, of course, is that this particular vinylidineamine has a collinear chain of atoms in the solid state, and we would not be justified in assuming that all vinylidineamines must have a similar configuration in all circumstances. The collinearity might be a consequence of the N-methyl group or of the electronegative

† P. J. Wheatley, *Acta crystallogr.* 1954, **7**, 68.

methylsulphonyl groups, and further work will be required to establish a complete picture of the stereochemistry of nitrogen in compounds of this type. However, the X-ray work has shown that, when nitrogen is joined to one atom by what is formally a double bond and to another atom by a single bond, the three atoms may, in certain circumstances, be collinear.

We have perhaps, in this chapter, emphasized the limitations and failures of classical stereochemical methods, instead of giving due credit to the many undoubted successes. We have done this because of the uncertainty attached to many chemical arguments. Nowadays an appeal is almost always made to the more reliable physico-chemical methods, particularly as the latter can frequently provide values of molecular parameters as well as disclose the molecular symmetry.

REFERENCE

J. C. BAILAR, *The Chemistry of Coordination Compounds* (Reinhold).

XI

DIPOLE MOMENTS

11.1. Dipole moments and symmetry

THE bond in a diatomic molecule can be truly covalent only
if the molecule is homonuclear. If the two atoms are not of the
same kind, the bond must have some ionic character because
the more electronegative atom will draw the bonding electrons
towards itself. We can describe this heteronuclear system by
placing a charge $-q$ on the more electronegative atom and a

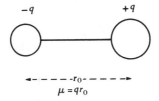

FIG. 1. The definition of a dipole moment.

charge $+q$ on the less electronegative atom (Fig. 1). If the
atoms are at a distance r_0 apart, the *electric dipole moment* is
defined as
$$\mu = qr_0.$$
Since q is of the order of one electron and the charge on an electron is 4.8×10^{-10} e.s.u., and since r_0 is of the order of 1 Å,

$$\mu \sim 4.8 \times 10^{-10} \times 10^{-8} \text{ e.s.u. cm}$$
$$\sim 4.8 \text{ Debye units}$$
$$\sim 4.8 \text{ D.}$$

This picture of the charge distribution in a molecule is highly
simplified. We have considered the charge centres as points,
which is clearly not true for real atoms. Moreover, we have
assumed that the only electrons giving rise to the dipole moment
are those involved in bonding, whereas lone pairs of electrons
may have moments equally as large as the bond moments.
Nevertheless, since we shall have little interest in the actual

magnitude of dipole moments, our simple picture will be sufficient for our purposes.

A bond moment is a vector quantity directed along the line of the internuclear axis. Provided we bear in mind the restrictions

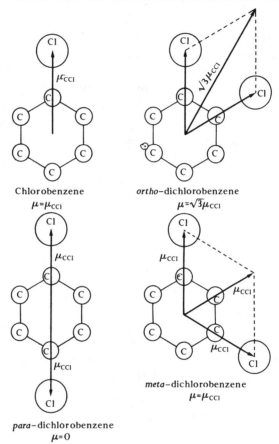

Chlorobenzene
$\mu = \mu_{CCl}$

ortho-dichlorobenzene
$\mu = \sqrt{3}\mu_{CCl}$

para-dichlorobenzene
$\mu = 0$

meta-dichlorobenzene
$\mu = \mu_{CCl}$

FIG. 2. The vector addition of bond dipole moments in dichlorobenzenes.

mentioned in the previous paragraph, we may approximate the total dipole moment of a polyatomic molecule by the vector sum of the various bond moments. Thus if $\mu(CCl)$ is the dipole moment of chlorobenzene, we would expect the moment of *ortho*-dichlorobenzene to be $\sqrt{3}\mu(CCl)$, that of *meta*-dichlorobenzene to be $\mu(CCl)$, and that of *para*-dichlorobenzene to be zero (Fig. 2).

This *vector addition rule* is by no means strictly valid, since both the magnitude and the direction of an individual bond moment may be altered by interaction with the other dipoles in the molecule. However, even though the vector sum of the individual bond moments may not give an accurate quantitative estimate of the total dipole moment, the latter, being a vector quantity, must depend on the molecular symmetry. Simple considerations of symmetry show that a molecule cannot have a dipole moment if it possesses a centre of symmetry, nor if it has symmetry C_{nh} or D_{nh} $(n > 1)$, i.e. a rotation axis normal to a mirror plane, nor if it has more than one axis of symmetry, nor if it has an alternating or inversion axis. The point group to which a molecule with a dipole moment might belong is, therefore, limited to the *polar point groups* shown in Table XLI.

TABLE XLI

The possible point groups to which a molecule with a finite dipole moment might belong

Schoenflies	Hermann–Mauguin
C_1	1
C_2	2
C_3	3
C_4	4
C_6	6
C_h	m
C_{2v}	mm
C_{3v}	$3m$
C_{4v}	$4mm$
C_{5v}	$5mm$
C_{6v}	$6mm$
.	.
.	.
.	.
$C_{\infty v}$	∞m

There are two main applications of dipole moment determinations. One involves the interpretation of the magnitudes of the moments in terms of the electronic distribution within the molecule. Such calculations are hampered by the various sorts of perturbations that occur in or between molecules, and are concerned with the more subtle aspects of molecular structure which we shall not consider in this book. The other application, as implied above, is to the determination of molecular symmetry.

If, as frequently happens, a decision about the molecular symmetry depends on whether the molecule has or has not a finite dipole moment, considerable confidence can be placed in the conclusion. If, however, the conclusion depends on the magnitude of the dipole moment, it must be treated with more caution. Thus two isomeric dichloroethylenes, which are both planar molecules (Fig. 3), have dipole moments of zero and 1·89 D.† There is then no doubt that the molecule with the

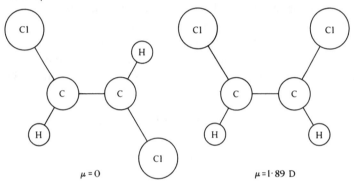

$\mu = 0$ $\mu = 1\cdot 89$ D

FIG. 3. The two isomers of dichloroethylene.

finite dipole moment must be the *cis* isomer with symmetry *mm* (C_{2v}), and that the molecule with the zero dipole moment must be the *trans* isomer with symmetry $2/m$ (C_{2h}). However, if we compare CH_3Cl, CH_2Cl_2, and $CHCl_3$, no certain conclusions about the shapes of these molecules can be reached. If we assume that these three molecules are tetrahedral, their dipole moments should be in the ratio $1:1\cdot155:1$, whereas if they are square planar and CH_2Cl_2 has the *cis* form, the ratio should be $1:1\cdot414:1$. The experimental values are 1·86, 1·60, and 1·1 D, which favour neither one nor the other configuration. In fact all three molecules are, of course, approximately tetrahedral.

11.2. Dipole moments and polarization

When an insulator is placed between the plates of a capacitor, the capacitance is increased by a factor ϵ known as the *dielectric*

† Values of dipole moments are taken from L. G. Wesson, *Tables of Electric Dipole Moments* (Massachusetts Institute of Technology Press).

constant. If c_0 is the capacitance with a vacuum and c the capacitance with the insulator between the plates of the capacitor,

$$\epsilon = \frac{c}{c_0}.$$

Since the introduction of any insulator increases the capacitance, and since the charge on the capacitor plates remains unchanged, the electric field between the plates must be reduced by the same

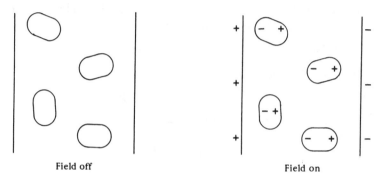

Field off   Field on

FIG. 4. The effect of an electric field on molecules without a permanent dipole moment.

factor ϵ. The reduction in the field may be due to two effects. In the first, which is always present whether the molecule of the insulator has a permanent dipole moment or not, a separation of the positive and negative charges within each molecule tends to take place (Fig. 4). The molecules are said to suffer a *distortion polarization* and an average dipole moment $\bar{m}$ is induced in the molecules. For isotropic materials the induced moment is proportional to the local intensity of the field F,

$$\bar{m} = \alpha_D F,$$

where α_D is called the *distortion polarizability*. The latter is the induced dipole moment per unit field strength and is a measure of the extent to which the electronic cloud of the molecule may be deformed. It has the dimensions of a volume, since the induced moment is of the form $q \times r$ and the field strength of the form q/r^2. The induced moment is independent of temperature since, if the position of a molecule is disturbed by a collision, a new dipole is immediately induced again in the direction of the field.

If, however, the molecule of the insulator has a permanent dipole moment μ, a second effect known as *orientation polarization* will be present, since the field tends to align the permanent dipoles along its own direction. The orientation polarization is temperature dependent and will decrease with an increase in temperature, since random thermal collisions will oppose the tendency of the permanent dipoles to orient themselves in the electric field. It can be shown that the *orientation polarizability* is given by

$$\alpha_O = \mu^2/3kT,$$

and hence the total polarizability α_T by

$$\alpha_T = \alpha_D + \alpha_O = \alpha_D + \mu^2/3kT.$$

The total polarization is given by the *Debye equation*

$$P_T = \left(\frac{4\pi n}{3}\right)(\alpha_D + \mu^2/3kT),$$

where n is the number of molecules per cm^3. If we multiply both sides of this equation by the ratio of the molecular weight M to the density d, we obtain the *total molar polarization*

$$P_M = P_D + P_O = \frac{4\pi N_0}{3}(\alpha_D + \mu^2/3kT),$$

where P_D and P_O are the *molar distortion polarization* and the *molar orientation polarization* respectively, and we have replaced nM/d by N_0.

If we restrict our considerations to gases and to dilute solutions of polar molecules in non-polar solvents, we can also express the total molar polarization by the *Clausius–Mosotti equation*

$$P_M = \left(\frac{\epsilon-1}{\epsilon+2}\right)\frac{M}{d}. \tag{1}$$

If we equate these two expressions for the total molar polarization, we obtain

$$\left(\frac{\epsilon-1}{\epsilon+2}\right)\frac{M}{d} = \frac{4\pi N_0}{3}(\alpha_D + \mu^2/3kT) = P_M. \tag{2}$$

11.3. Determination of dipole moments by the temperature method

In order to obtain a value for the dipole moment of a molecule in the gas phase or in dilute solution in a non-polar solvent, we can measure the dielectric constant over a range of temperature and plot P_{M}, calculated from equation (1), against $1/T$. From equation (2) the slope of the graph will be equal to $4\pi N_0\,\mu^2/9k$, and the intercept at $1/T = 0$ will give the molar distortion polarization P_{D}. For gases ϵ is close to unity and equation (2) may be simplified to give

$$\left(\frac{\epsilon-1}{3}\right)\frac{M}{d} = \frac{4\pi N_0}{3}(\alpha_{\mathrm{D}}+\mu^2/3kT) = P_{\mathrm{M}}.$$

The density of a gas is normally calculated from the gas laws. Departure from ideal behaviour may, if necessary, be eliminated by measurement of the dielectric constant over a range of pressure at each temperature followed by extrapolation of the total molar polarization to zero pressure. The density of a solution may be found by means of a pyknometer. Dielectric constants are found by measurement of the capacitance of a condenser which contains the gas or solution between the plates.

We may illustrate a typical application of the temperature method by considering the determination of the dipole moment of bromine pentafluoride BrF_5.† The dielectric constant of the vapour was measured at seven different temperatures. The results are shown in the first two columns of Table XLII. At each temperature the value of the total molar polarization may be calculated from equation (1). Thus at $T = 345\cdot6°$ K,

$$P_{\mathrm{M}} = \frac{0\cdot006320\times22421\times345\cdot6}{3\times273\cdot2}$$

$$= 59\cdot6 \text{ cm}^3.$$

The calculated values of P_{M} at each temperature are shown in the

† M. T. Rogers, R. D. Pruett, H. B. Thompson, and J. L. Spiers, *J. Am. chem. Soc.* 1956, **78**, 44.

last column of Table XLII. P_M is then plotted against $1/T$ as shown in Fig. 5. From the slope of the graph we find

$$\frac{4\pi N_0 \mu^2}{9k} = \frac{10}{0 \cdot 002916 - 0 \cdot 002211}.$$

TABLE XLII

The dielectric constant and total molar polarization of BrF_5 *as a function of temperature*

Temp.	ϵ	$1/T$	P_M
345·6° K	1·006320	0·002894	59·6 c.c.
362·6	1·005824	0·002758	57·7
374·9	1·005525	0·002667	56·6
388·9	1·005180	0·002571	55·0
402·4	1·004910	0·002485	54·0
417·2	1·004603	0·002397	52·5
430·8	1·004378	0·002321	51·5

Therefore
$$\mu^2 = \frac{90 \times 1 \cdot 380 \times 10^{-16} \times 10^{-23}}{4 \times 3 \cdot 142 \times 0 \cdot 000705 \times 6 \cdot 025}$$
$$= 2 \cdot 327 \times 10^{-36}.$$

Hence
$$\mu = 1 \cdot 52 \times 10^{-18} \text{ e.s.u. cm}$$
$$= 1 \cdot 52 \text{ D.}$$

An extrapolation of the graph to infinite temperature yields a value of $21 \cdot 8$ cm^3 for the molar distortion polarization.

The fact that the molecule possesses a finite dipole moment means that its possible symmetry is greatly restricted. BrF_5 cannot be trigonal bipyramidal, nor can it be pentagonal planar. If it is assumed that $\mu(BrF)$ is the same in BrF_5 as it is in BrF ($1 \cdot 29$ D), the most plausible structure for the molecule is a square-pyramidal arrangement of symmetry $4mm$ (C_{4v}) with the bromine atom slightly below the plane of the four fluorine atoms that form the base of the pyramid (Fig. 6).

11.4. The refractivity method

The temperature method for the determination of dipole moments gives reliable results, except for some anomalous cases that we shall mention in section 11.6. However, not all sub-

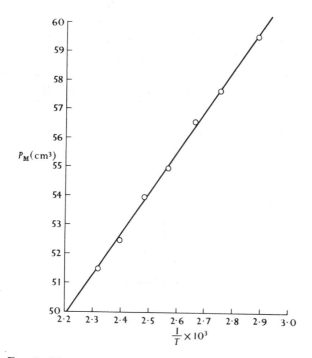

Fig. 5. Plot of the total molar polarization of BrF_5 against the reciprocal of the absolute temperature.

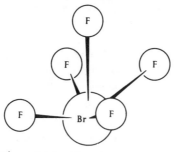

Fig. 6. The molecule of bromine pentafluoride, BrF_5.

stances are stable over a sufficiently wide range of temperature for this method to be applied. A second method, known as the *refractivity method*, has, therefore, been devised to eliminate the effect of the distortion polarization. The distortion polarization arises in two ways. Firstly, the electrons in the sample molecules will be displaced with respect to the nuclei towards the positive pole of the applied electric field. This effect is known as the *electron polarization* P_E. Secondly, the nuclei will be slightly displaced with respect to one another. This effect

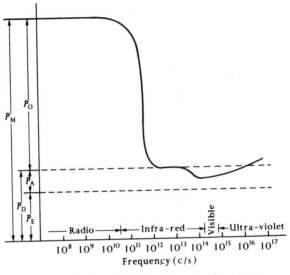

FIG. 7. Dependence of the polarization of a molecule on the frequency of the applied field. (Adapted from N. V. Sidgwick, *The Covalent Link in Chemistry*.)

is called the *atom polarization* P_A. Thus $P_D = P_E + P_A$. The refractivity method takes advantage of the fact that the speed of response of these three types of polarization (electron, atom, and orientation) to an oscillatory electric field differs greatly. If we plot polarization against the frequency of an alternating electric field, we obtain the graph shown in Fig. 7. The electron polarization is brought about by fields up to very high frequencies. The atom polarization, which is generally much smaller than the electron polarization, responds less readily. The reorientation of a polar molecule is a much slower process still. Consequently if we measure the dielectric constant at a low frequency and at a high frequency, the difference, after we have allowed for the atom polarization, should give us the orientation polarization from which we can calculate the dipole moment as before. An additional simplification can be introduced through the relation that the dielectric constant is equal to the square of the refractive index, provided the refractive index is measured at sufficiently high frequencies for the orientation polarization to be absent. Thus

$$P_E = \left(\frac{n_r^2-1}{n_r^2+2}\right)\frac{M}{d}, \qquad (3)$$

where n_r is the refractive index measured under suitable conditions. P_E is often called the *molar refractivity* as well as the molar electron polarization. The refractivity method, therefore, involves the determination of the dielectric constant at radio frequencies and of the refractive index at visible frequencies. If a gas is used as the sample, it is best to make a series of readings at different pressures and to extrapolate to zero pressure in order to eliminate the effect of interaction between the polar molecules. If the sample is dissolved in a non-polar solvent, interaction between the solute molecules may be eliminated by extrapolation of a series of measurements taken at different concentrations to infinite dilution. However, in the latter case it is not possible to allow experimentally for any interaction between solute and solvent molecules, and the presence of this *solvent effect* may sometimes be serious (section 11.6). It is also necessary to allow for the atom polarization P_A. We saw in the last section that $P_D = P_E + P_A$ can be obtained by the temperature method. Since P_E can be determined from refractive index measurements, P_A may be evaluated. For substances which cannot be investigated by the temperature method an estimate of P_A may be made from measurements on a similar compound which is stable over a sufficient range of temperature. It is found that P_A usually lies between 5 per cent and 15 per cent of P_E.

As a typical illustration of the use of the refractivity method we may consider the determination of the dipole moment of bromine trifluoride, BrF_3, which is stable in the vapour phase over only a limited temperature range and which is too reactive to be dissolved in a non-polar solvent.† The dielectric constant was measured by the *heterodyne beat method*. In this method there are essentially two electrical circuits. One consists of a crystal-controlled oscillator of known frequency (500 kc/s). The other circuit is of variable frequency and consists of a variable frequency oscillator, a cell-capacitor into which the sample can be introduced, and a precision capacitor that can be adjusted to alter the frequency of the circuit. At the start of the experiment the cell-capacitor is evacuated and the variable frequency circuit is made to beat with the fixed frequency circuit, the beat frequency being compared with a second fixed frequency (400 c/s) on an oscilloscope. The sample gas is then introduced into the cell-capacitor with a resulting increase in the capacity of the variable frequency circuit. The beat frequency changes and is restored to its original value by adjustment of the precision capacitor. The cell-capacitor is calibrated with a gas of known dielectric constant, usually ammonia or carbon dioxide, and the capacitance of the cell and its leads calculated from the dielectric constant of the calibrating gas.

With BrF_3 as the sample gas, the capacitance of the cell was measured in this way at several gas pressures at a given temperature. A plot of capacitance against pressure gave the change in capacitance from vacuum to one atmosphere. In order to eliminate errors arising from deviations from the gas laws and from adsorption of the gas on the walls of the cell-capacitor, only the straight line portion of the graph was considered. From the change in capacitance and from the known capacitance of the cell the dielectric constant of BrF_3 at one atmosphere was calculated. Measurements were repeated at three different temperatures and the total molar polarization P_M at each temperature was calculated from equation (1). The refractive index was measured with sodium light and the molar refractivity calculated from equation (3). P_E was found to be 12·92 cm³. The molar atom polarization P_A of ClF_3 has been estimated with

† M. T. Rogers, R. D. Pruett, and J. L. Spiers, *J. Am. chem. Soc.* 1955, **77**, 5280.

some care to be 5·5 cm³, and it was assumed that the value for BrF₃ would be the same. The molar orientation polarization P_O is given by the difference between P_M and P_D. P_O was computed at each temperature, and the dipole moment obtained from

$$\mu = \sqrt{(9kTP_O/4\pi N_0)}$$
$$= 0\cdot0128\sqrt{(P_O\,T)}.$$

The full results are shown in Table XLIII. The mean dipole moment was found to be 1·19 D, which is consistent with a planar T-shaped molecule like ClF_3 of *mm* (C_{2v}) symmetry (Fig. 8).

TABLE XLIII
The dipole moment of BrF₃ *determined by the refractivity method*

Temp.	ϵ	P_M	$P_D = P_E + P_A$	$P_O = P_M - P_D$	μ
415·5° K	1·003748	42·6 cm³	18·42 cm³	24·18 cm³	1·28 D
424·7	1·003189	37·0	18·42	18·58	1·15
448·2	1·002964	36·3	18·42	17·88	1·14
				Mean .	1·19 D

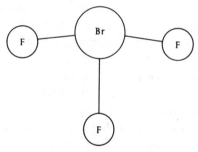

FIG. 8. The molecule of bromine trifluoride, BrF₃.

The refractivity method is widely used because of its rapidity and more general applicability. The results, however, are rather less reliable than those obtained by the temperature method.

11.5. Other methods used for the determination of dipole moments

The most accurate method for the determination of dipole

moments is microwave spectroscopy. If an electric field is applied to a gas, the pure rotational lines are split into their *Stark components*, and the amount of splitting is a function of the electric field strength and of the dipole moment. The Stark effect in an electric field is entirely analogous to the Zeeman effect in a magnetic field, and the splitting arises in both effects because the spatial degeneracy of the energy levels is removed by the applied electric or magnetic field. The individual Stark components can be resolved by fields of a few thousand V/cm and the splitting can be measured with great accuracy. The electric field strength is usually determined by calibration with molecules of known dipole moment. Since the sample is in the vapour state and at a low pressure, there will be no solvent effect, and the interaction between the polar molecules is minimized. The presence of impurities will not affect the results provided the composite spectrum can be analysed. Moreover, in favourable cases, the dipole moment of each isotopic species in each separate vibrational state can be obtained. The method is restricted to simple molecules with high vapour pressures, but there is already available a considerable amount of accurate quantitative information on molecular dipole moments which should be capable of interpretation in terms of the detailed electronic structure of the molecules.

More comprehensive but less accurate information may be obtained from infra-red vibration-rotation spectra. We have already seen (section 2.7) that a vibration will be inactive in the infra-red unless it involves a change in the dipole moment of the molecule. The greater the change in dipole moment the more intense will be the infra-red absorption. Thus the absolute intensity of an infra-red band can be related to the dipole moment, not of the whole molecule, but of that portion of the molecule which is responsible for the interaction with the incident radiation. In this way bond dipole moments of both polar and non-polar molecules may be found, although it is difficult to be sure of the precise significance of a bond moment determined by this method. The measurement of absolute intensities is not easy and the interpretation of the results must necessarily be made in terms of the displacement of the electrons during the vibration, so that progress has been slow. Nevertheless, the study of absolute infra-red intensities appears to be a promising method for the elucidation of the electronic structure of molecules.

Finally, the molecular beam method is useful for substances of low volatility or solubility. In the original method the solid polar compound was heated in an oven so as to give a detectable vapour pressure, and a narrow beam of molecules was allowed to pass through a slit system into a vacuum. The beam then traversed a highly inhomogeneous electric field and was condensed on a cool metal plate. In the absence of the electric field, the beam leaves a vertical trace on the receiver. When the field is switched on the trace shifts, since a small dipole is induced in the molecules. If the molecules are polar, the trace not only shifts but broadens, since the amount of shift of a polar molecule depends on the inclination of the permanent dipole to the electric field. The broadening is proportional to the dipole moment. The apparatus must be calibrated with a substance of known dipole moment. Values of dipole moments obtained in this way are not very accurate, but the results originally obtained were sufficiently reliable to show that some of the alkali halides had moments of the order of 10 D.

The molecular beam method has been refined into what is known as the *electric resonance method*. This method not only yields reliable values for dipole moments but gives extremely accurate values for equilibrium internuclear distances. The beam is passed through two inhomogeneous electric fields, the first of which deflects the beam and the second refocuses it onto a tungsten wire detector. Between these two fields is a third, homogeneous, electric field and on this third field is superimposed an oscillatory field perpendicular to the static homogeneous field. The oscillatory field, which has a frequency in the microwave region, induces transitions between the rotational levels of the polar molecules. At the resonance frequency a certain portion of the polar molecules are in an excited state and they are no longer focused onto the detector by the second inhomogeneous field.

Consequently under electric resonance conditions the detected beam shows a marked decrease in intensity. From the frequency of the oscillatory field and from the magnitude of the static homogeneous field it is possible to obtain a value for $\mu^2 I_e$, where I_e is the equilibrium moment of inertia. By observation of the dependence of the resonance frequency on applied field strength, a value for μI_e may be found. Thus both μ and I_e may be evaluated, and from I_e a value of r_e for a diatomic species may be calculated. Some results obtained by the electric resonance method are shown in Table XLIV.

TABLE XLIV

Values obtained by electric resonance of the equilibrium inter-nuclear distances and dipole moments of vaporized salts†

Molecule	r_e	Dipole moment
KF	2·55 Å	7·33 D
KCl	2·6666	10·4
KBr	2·8207	10·5
CsF	2·34	7·88
CsCl	2·88	10·46
TlCl	2·541	4·444

11.6. Applications of dipole moment determinations

The dipole moments of most simple molecules have been determined, and the molecular symmetries deduced from these moments are usually consistent with the structures found by other methods. It should be noticed that none of the methods discussed above gives the absolute direction of the dipole moment, so that it is not possible experimentally to distinguish the positive from the negative end of a dipole. A list of the dipole moments and the symmetries of some simple molecules is given in Table XLV. Occasionally finite values are obtained when a zero value is expected, but in such cases it is usually found that a particularly strong solvent effect exists. An example is afforded by the dipole moment of mercuric chloride, $HgCl_2$, in dioxane, $C_4H_8O_2$. There is plenty of evidence that $HgCl_2$ is linear with

† Extracted from N. F. Ramsey, *Molecular Beams* (Clarendon Press).

∞/mm $(D_{\infty h})$ symmetry and should thus have a zero dipole moment. A value of 1·29 D was found by the temperature method with dioxane as a solvent.† More recent work has shown that dioxane forms addition compounds of the oxonium type with a wide variety of substances, so that the finite moment is not surprising.

TABLE XLV

Dipole moments and symmetries of some simple molecules

Molecule	Dipole moment	Symmetry
NO	0·16 D	$\infty m\,(C_{\infty v})$
CO	0·12	$\infty m\,(C_{\infty v})$
Cl_2	0	$\infty/mm\,(D_{\infty h})$
H_2O	1·86	$mm\,(C_{2v})$
CO_2	0	$\infty/mm\,(D_{\infty h})$
BCl_3	0	$\bar{6}2m\,(D_{3h})$
NH_3	1·47	$3m(C_{3v})$
$SiCl_4$	0	$\bar{4}3m\,(T_d)$
SF_6	0	$m3m\,(O_h)$

Tables of bond moments can be drawn up although, as we saw in section 11.1, the values cannot be considered to be very reliable and must be used with caution. These bond moments have been used to estimate bond angles. For instance, if we assume that $\mu(CCl)$ is equal to the dipole moment of CH_3Cl (1·86 D), then the observed dipole moment of $CHCl_3$ (1·1 D) requires a ClHCl angle of 116°. This agrees moderately well with the value of 110° 24′ obtained by microwave spectroscopy. On the other hand, the observed moment of CH_2Cl_2 (1·60 D) requires a ClHCl angle of 130°, which may be compared with the microwave value of 111° 47′. There are numerous examples of these anomalies, and we can only conclude that the use of bond moments may or may not give reasonable values for interbond angles.

Dipole moment determinations are especially valuable for the elucidation of the symmetry of large organic molecules which cannot readily be studied by other methods. For example,

† W. J. Curran and H. H. Wenzke, *J. Am. chem. Soc.* 1935, **57**, 2162.

thianthren, $C_{12}H_8S_2$, and the corresponding selenium and tellurium compounds have finite moments, whereas phenazine, $C_{12}H_8N_2$, has a zero moment. Thus it is probable that phenazine is planar (Fig. 9(a)), whereas thianthren is folded across the S...S line (Fig. 9(b)). These tentative conclusions were confirmed

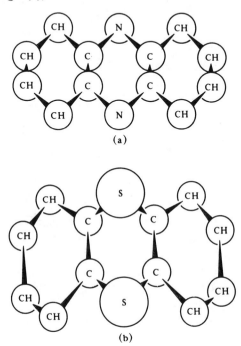

(a)

(b)

FIG. 9. Planar and folded heterocyclic molecules: (a) phenazine, (b) thianthren.

many years later by X-ray diffraction studies of the solids.†
However, it must be pointed out that the possibility of a solvent effect always leaves a suspicion that a dipole moment that is observed to be finite should really have been zero, and most workers prefer to have conclusions drawn from dipole moments confirmed by another method.

Interesting information may also be gained about rotation around single bonds. A substance like 1, 2-dichloroethane,

† F. L. Hirshfeld and G. M. J. Schmidt, *J. chem. Phys.* 1957, **26**, 923.
H. Lynton and E. G. Cox, *J. chem. Soc.* 1956, 4886.

$ClH_2C.CH_2Cl$, may exist in four possible forms. In the first place it may exist as molecules in which each half is rotating freely about the C—C bond. In this case the substance will have a finite dipole moment which is independent of temperature, and it can be shown that the mean moment of the molecules will be

$$\bar{\mu} = \sqrt{2}\sin\theta\,\mu(CCl),$$

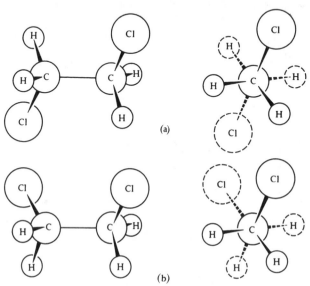

(a)

(b)

FIG. 10. Side and end views of the molecules of 1, 2-dichloroethane: (a) *trans* form of $2/m$ (C_{2h}) symmetry, (b) *gauche* form of $2(C_2)$ symmetry.

where $\mu(CCl)$ is the bond moment of the C—Cl bond and θ is the CCCl angle. If we take $\theta = 110°$ and $\mu(CCl) = 1\cdot86$ D, then $\bar{\mu} = 2\cdot47$ D. Secondly, the molecules may be fixed permanently in the *trans* configuration (Fig. 10 (a)) of $2/m$ (C_{2h}) symmetry, in which case the substance will have zero dipole moment at all temperatures. Thirdly, the molecules may have the *gauche* configuration (Fig. 10 (b)) of 2 (C_2) symmetry, in which case the substance will have a finite, temperature-independent dipole moment. Finally, if the barrier to interconversion is not too high, the substance may exist as an equilibrium mixture of *gauche* and *trans* molecules. In these circumstances there will tend to be more molecules of the form with lower potential energy, that is the *trans* form. Since this is the form with zero dipole moment, it follows that the observed dipole moment for hindered rotation should be less than that for free rotation, and also that, as the temperature is raised and more *gauche* molecules are formed, the observed dipole moment should increase. Clearly an investigation of the dipole moment and its temperature

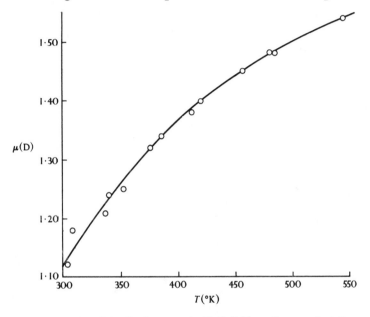

FIG. 11. Plot of the dipole moment of 1, 2-dichloroethane against the absolute temperature.

dependence should be able to distinguish between these four possibilities.

Two sets of workers have measured the dipole moment of 1, 2-dichloroethane as a function of temperature and their results are presented together in Fig. 11.† The form of the curve leaves no doubt that the substance exists as a mixture of *gauche* and *trans* isomers, and from the temperature dependence of

† (a) C. T. Zahn, *Phys. Rev.* 1931, **38**, 521. (b) I. Watanabe, S. Mizushima, and Y. Morino, *Sci. Pap. Inst. phys. chem. Res., Tokyo* 1942, **39**, 401.

the dipole moment it is possible to deduce that there is an energy difference of 1·2 kcal/mole between the two isomers.

REFERENCES

P. DEBYE, *Polar Molecules* (The Chemical Catalog Co.).

R. J. W. LE FEVRE, *Dipole Moments* (Methuen).

J. W. SMITH, *Electric Dipole Moments* (Butterworth).

S. MIZUSHIMA, *Structure of Molecules and Internal Rotation* (Academic Press).

XII

MAGNETIC METHODS

12.1. Magnetic susceptibilities

IF a substance is placed in a magnetic field, the *magnetic induction* or the total number of lines of force per unit area is given by

$$B = H + 4\pi I,$$

where H is the field strength in a vacuum and I is the *intensity of magnetization* or magnetic moment per unit volume. We may define a constant ϵ' known as the *permeability*, which is the magnetic analogue of the dielectric constant ϵ, such that

$$\epsilon' = \frac{B}{H}.$$

In a vacuum $B = H$ and $\epsilon' = 1$. In magnetic experiments it is usually more convenient to work in terms of I/H rather than B/H, and we define

$$\kappa = \frac{I}{H},$$

where κ is the *volume susceptibility*. We may continue our analogy with electrical polarization, and multiply κ by the ratio of the molecular weight to the density and so obtain the *molar susceptibility* χ_M,

$$\chi_M = \frac{M\kappa}{d}.$$

In a similar way, we may express our results in terms of the gramme-atomic susceptibility χ_A or the gramme-ionic susceptibility χ_I. The molar susceptibility is given by an equation similar to that for the molar polarization (section 11.2):

$$\chi_M = \chi_D + \chi_P = N_0\left(\alpha_D + \frac{\mu_M^2}{3kT}\right), \tag{1}$$

where α_D is the *induced magnetic susceptibility* per molecule and μ_M is the permanent magnetic moment of a molecule. χ_D and χ_P are called the *molar diamagnetic susceptibility* and the *molar paramagnetic susceptibility* respectively.

At this point we must drop our analogy between magnetic and electrical effects since χ_D and χ_P have opposite signs, whereas the corresponding electrical terms (the distortion polarization P_D and the orientation polarization P_0) have the same sign. If our substance has no permanent magnetic moment, χ_P is zero and the only term which contributes to the molar susceptibility is the negative term χ_D. The substance is then said to be *diamagnetic*. It is less permeable to magnetic lines of force than is a vacuum, and will tend to move from a stronger to a weaker part of an inhomogeneous magnetic field. If, however, our substance possesses permanent magnetic dipoles, the positive χ_P term overshadows the negative χ_D term, and the molar susceptibility χ_M is positive. The substance is then said to be *paramagnetic*. It is more permeable to magnetic lines of force than is a vacuum, and will tend to move from a weaker to a stronger part of the magnetic field. For a few substances, mainly metals and alloys, χ_M is positive and numerically about a million times larger than normal diamagnetic susceptibilities. These substances are called *ferromagnetic* (section 9.5). We shall not be concerned with ferromagnetic substances in this chapter.

Diamagnetism, like distortion polarization, is exhibited by all substances and is a relatively weak effect. As we have seen, paramagnetism is always superposed on a diamagnetic effect, but the diamagnetic susceptibility may be eliminated by measurement of the molar susceptibility over a range of temperatures. Since the diamagnetic effect is only about one-tenth of the total susceptibility, it is more common, however, to make a diamagnetic correction by the method we shall discuss in section 12.3, rather than to make use of a temperature method.

When the effects of diamagnetism have been eliminated, we may modify equation (1) and write

$$\chi_P = \frac{N_0 \mu_M^2}{3kT}. \tag{2}$$

This equation shows that the paramagnetic contribution to the total susceptibility should be inversely proportional to the absolute temperature. This is true of a number of compounds, which are then said to obey *Curie's law.* It is more usual, however, for paramagnetic substances to obey a relation of the form

$$\chi_P \propto \frac{1}{T+\theta},$$

which is known as the *Curie–Weiss law.* θ is a constant over a restricted range of temperatures and a value for θ may be obtained from the variation of the magnetic susceptibility with temperature. The breakdown of the Curie law is due to the effect of neighbouring paramagnetic atoms, so that we would not expect the Curie law to be obeyed in the solid state or in concentrated solutions, where the density of paramagnetic atoms will be high. Nevertheless, since θ is seldom known, we often pretend that the Curie law is obeyed, and use equation (2) to obtain a value for the magnetic moment. We then acknowledge this approximation by replacing μ_M by μ_{eff}, the *effective magnetic moment.* Thus we obtain from equation (2)

$$\mu_{eff} = (3kT\chi_P/N_0)^{\frac{1}{2}},$$

and we must not be too surprised if agreement between μ_{eff} and a predicted magnetic moment is not good.

12.2. Measurement of magnetic susceptibilities

A convenient apparatus for the measurement of susceptibilities, provided several grammes of substance are available, is Gouy's magnetic balance. A sample of the material, contained in a glass cylinder of internal cross-section A, is suspended from a sensitive balance. One end of the sample cylinder lies between the poles of a magnet capable of producing a strong magnetic field H_1 of about 10,000 gauss. At the other end of the cylinder the magnetic field strength H_2 is often negligible compared with H_1. If the difference between the apparent mass with and without the field is found to be Δm, the force acting on the specimen because of the magnetic field gradient is

$$F = \Delta m g,$$

where g is the gravitational constant. If the volume susceptibility of the surrounding medium, which is usually air, is κ_2 and that of the sample is κ_1, then

$$F = \tfrac{1}{2}(\kappa_1 - \kappa_2)(H_1^2 - H_2^2)A.$$

If H_2 is negligible compared with H_1,

$$\Delta m\,g = \tfrac{1}{2}(\kappa_1 - \kappa_2)H_1^2 A.$$

Thus $\qquad\qquad \kappa_1 = \kappa_2 + 2\Delta m\,g/A H_1^2.$

From the volume susceptibility of the sample κ_1 we may obtain the molar susceptibility by multiplying by the molecular weight and dividing by the density. The measurement of the dimensions of the sample tube and of the field strength may be avoided if the apparatus is calibrated with a material of known susceptibility, such as a solution of nickel chloride.

12.3. Diamagnetism

Since an electron in an orbital produces a magnetic field which is equivalent to that of an electron moving in a circle, the application of an external magnetic field will give rise to a magnetic field opposing the external field. The susceptibility per gramme-atom of a material placed in an external field can be shown to be given by

$$\chi_{\text{A}} = -\frac{N_0 e^2}{6 m_{\text{e}} c^2} \sum_{i=1}^{N} \overline{r_i^2}, \tag{3}$$

where N is the number of electrons in the atom, $\overline{r_i^2}$ is the mean square radius of the orbit of the ith electron, m_{e} is the electron mass and N_0, e, and c are fundamental constants. Diamagnetism, is, therefore, independent of temperature and is a function of the size of an atom. We would expect the gramme-atomic susceptibility to vary directly with atomic radius. Table XLVI shows that this conclusion is justified.

Diamagnetic susceptibilities are approximately additive, provided that a *constitutive correction* is made for the presence of multiple bonds, ring closure, and certain other anomalous effects, some of which are shown in Table XLVII. The molar diamagnetic susceptibility of a molecule with N atoms may be

TABLE XLVI

Gramme-atomic and gramme-ionic susceptibilities compared with atomic and ionic radii

Atom	$\chi_A \times 10^6$	Tetrahedral radius in crystals	Ion	$\chi_I \times 10^6$	Ionic radius in crystals
F	$-11\cdot5$ cm³	$0\cdot64$ Å	F⁻	$-13\cdot9$ cm³	$1\cdot33$ Å
Cl	$-20\cdot1$	$0\cdot99$	Cl⁻	$-24\cdot1$	$1\cdot81$
Br	$-30\cdot6$	$1\cdot11$	Br⁻	$-34\cdot6$	$1\cdot95$
I	$-44\cdot6$	$1\cdot28$	I⁻	$-48\cdot6$	$2\cdot16$
S	$-15\cdot0$	$1\cdot04$	Li⁺	$-0\cdot2$	$0\cdot60$
Se	$-23\cdot0$	$1\cdot14$	Na⁺	$-5\cdot2$	$0\cdot95$
Te	$-37\cdot3$	$1\cdot32$	K⁺	$-14\cdot5$	$1\cdot33$

TABLE XLVII

Molar constitutive correction constants†

Molecular features	$\lambda \times 10^6$
C=C	$+5\cdot5$ cm³
C≡C	$+0\cdot8$
N=N	$+1\cdot8$
C=N	$+8\cdot2$
C—Cl	$+3\cdot1$
C—Br	$+4\cdot1$
C—I	$+4\cdot1$
C in one aromatic ring . .	$-0\cdot24$
C in two aromatic rings . .	$-3\cdot1$
C in three aromatic rings .	$-4\cdot0$

calculated from lists of such constants by means of the formula

$$\chi_D = \sum_{i=1}^{N} n_i \chi_i + \lambda,$$

where n_i and χ_i are the number and the atomic susceptibilities of the ith atom, and λ is the constitutive correction. In this way a diamagnetic correction may be applied to a paramagnetic susceptibility measured at only one temperature.

In crystals it is often found that the diamagnetic susceptibility is anisotropic, and the determination of the principal suscepti- bilities may be used to obtain information about the crystal structure. The method is mainly restricted to flat aromatic

† Extracted from E. A. Braude and F. C. Nachod, *Determination of Organic Structures by Physical Methods* (Academic Press).

molecules. In molecules of this sort we have to place a rather different interpretation on the meaning of r_i in equation (3). In an aromatic system the π electrons are no longer localized on one atom but occupy a molecular orbital which may encompass the whole molecule. r_i is thus no longer of the order of magnitude of an atomic radius, but of a molecular radius. Consequently, we would expect the magnetic susceptibility of an aromatic molecule to have normal values in the plane of the ring, but much larger values perpendicular to the molecular plane. This is confirmed experimentally. In naphthalene,† for example, the two principal susceptibilities in the ring are $-56 \cdot 1$ and $-53 \cdot 9 \times 10^{-6}$ cm³, whereas perpendicular to the ring the molar susceptibility is $-169 \cdot 0 \times 10^{-6}$ cm³. In the limiting case of graphite,‡ which is a giant molecule in two dimensions, the two principal susceptibilities in the molecular plane are -5×10^{-6} cm³ whereas perpendicular to this plane the value is -275×10^{-6} cm³.

Since we have this large susceptibility perpendicular to the aromatic ring, we can clearly reverse our argument and, by measuring the direction of maximum susceptibility with respect to the crystal axes, we may deduce the orientation of the molecular plane. This method has been used to confirm the orientation of the hexamethylbenzene molecule which crystallizes in the triclinic system with only one molecule in the unit cell. The molecules lie in a cleavage plane and the molar susceptibility perpendicular to this plane was found to be $-163 \cdot 8 \times 10^{-6}$ cm³. In the molecular plane the susceptibility is about -102×10^{-6} cm³.§ The method is unlikely to achieve any real importance in structure determinations, since its value usually decreases as the symmetry of the crystal increases. The reason for this is that, in the triclinic system, the planes of the molecules in the unit cell are almost always parallel. In crystal systems of higher symmetry the planes of the molecules in the unit cell are almost always inclined to each other, with the result that the direction of maximum susceptibility in the crystal is no longer

† K. S. Krishnan, B. C. Guha, and S. Banerjee, *Phil. Trans. R. Soc.* A 1933, **231**, 235.

‡ N. Ganguli and K. S. Krishnan, *Proc. R. Soc.* A 1941, **177**, 168.

§ K. Lonsdale and K. S. Krishnan, ibid. 1936, **156**, 597.

perpendicular to any one molecular plane. Magnetic suscepti-bility measurements can give useful information for all crystal systems except cubic, but very often the same information can be gained either by measurement of the refractive indices, or from a qualitative survey of the intensities of a diffracted beam of X-rays, or merely from the unit cell dimensions.

Before we leave diamagnetic susceptibilities one curious point should be noticed. The principal susceptibilities in the plane of the hexamethylbenzene molecule are not quite equal, and the same is true for all fully substituted benzene derivatives so far studied. No explanation of this apparent lack of magnetic symmetry has yet been advanced.

12.4. Paramagnetism

Paramagnetism is usually associated with the presence of unpaired electrons. A number of simple molecules such as NO, NO_2, and ClO_2 are paramagnetic because they possess an odd number of electrons. O_2 and S_2 are paramagnetic for a rather different reason. In these molecules the highest occupied molecular orbital is doubly-degenerate. Since there are two electrons to be placed in this doubly-degenerate orbital, Hund's Rule of Maximum Multiplicity requires one electron to be in each of the two equivalent orbitals. Thus O_2 and S_2 have two unpaired electrons, and are more strongly paramagnetic than the odd electron molecules mentioned above.[†] The absence of paramagnetism can be a useful guide to the correct formulation of a compound. For instance, all known salts of hypophosphoric acid, H_2PO_3, are diamagnetic, so that the correct formula for the acid is probably $H_4P_2O_6$ with all the electrons paired. For the same reason, dithionites are derivatives of $H_2S_2O_4$ rather than of HSO_2, and mercurous chloride should be written Hg_2Cl_2.

For our purposes a much more interesting cause of paramag-netism is the existence of partially filled inner electron shells in atoms and ions of the transition elements, the rare earths, and the actinides. An unpaired electron has both a spin angular

† For a full account of this topic see C. A. Coulson, *Valence* (Clarendon Press), Chapter 4.

momentum and an orbital angular momentum. The first of these may be regarded as the result of the electron's spinning about its own axis, and the second as the result of the motion of the electron around the atomic nucleus. It is a combination of these two effects that gives rise to the magnetic moment and hence to the paramagnetism. Provided we restrict our considerations to the first transition series, we can usually neglect the orbital contribution, and the magnetic moment then arises solely from the spin of the unpaired electrons. If there are n unpaired electrons, the magnetic moment is given by the 'spin-only' formula

$$\mu_{\text{eff}} = \sqrt{\{n(n+2)\}}. \tag{4}$$

In this equation μ_{eff} is expressed in terms of the Bohr magneton (B.M.) which is the natural unit of magnetism and is equal to the magnetic moment of an electron spinning on its own axis. 1 B.M. is equal to

$$eh/4\pi m_e c = 9\cdot 2837 \times 10^{-21} \text{ erg/gauss}, \tag{5}$$

where m_e is the mass of the electron and the other symbols have their usual meaning. Values of effective magnetic moments in B.M. calculated from equation (4) are shown in Table XLVIII.

TABLE XLVIII

Effective magnetic moment in terms of the number of unpaired electrons

Number of unpaired electrons (n)	1	2	3	4	5
μ_{eff} (B.M.)	1·73	2·83	3·87	4·90	5·92

If the number of unpaired electrons can be deduced from the observed magnetic moment, it is sometimes possible to obtain information about the valency of the paramagnetic atom, its stereochemistry, or the type of bond in a metallic complex.

As an illustration of the use of paramagnetism to decide questions of valency we may consider the free ferric and ferrous ions. The ferric ion has five $3d$ electrons and, according to Hund's rule, each electron will occupy singly one of the five available $3d$ orbitals with all five spins parallel. The magnetic moment will thus be about 5·9 B.M. On the other hand, the ferrous ion with six $3d$ electrons must have two of these paired in one of the $3d$

orbitals. Consequently, there will be only four unpaired electrons and the magnetic moment will be about 4·9 B.M. In a similar way, cuprous and cupric ions may be distinguished, since the latter is paramagnetic with a moment corresponding to one unpaired electron, whereas the cuprous ion is diamagnetic with all the electrons paired. If a valency is ascribed to a metal in the first transition series such that an odd number of electrons should

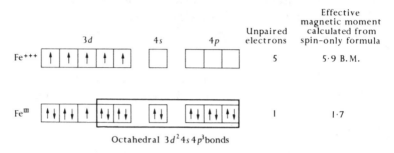

Fig. 1. The electronic structure of trivalent iron when free and octahedrally coordinated.

be present, and yet no paramagnetism is observed, then either the wrong valency has been chosen or some unsuspected form of bonding has taken place. We shall return to this second point later in this section. For heavier elements interaction between orbital and spin angular momenta and the breakdown of Hund's rule can make the magnetic criterion of valency less certain.

Much discussion has centred around the type of bond that exists in certain metallic complexes. In some octahedral ferric complexes a magnetic moment of about 5·9 B.M. is observed; in others the moment is only about 2·3 B.M., which is slightly greater than the value expected for one unpaired electron. It is clear that there must be a significant difference in the bonding in these two types of complex. At first it was suggested that the former were ionic compounds with the ferric ion left essentially free with its five unpaired electrons, whereas the latter were covalent compounds involving d^2sp^3 hybrid orbitals, as shown in Fig. 1. The fact that K_3FeF_6 with the highly electronegative fluorine atoms has a magnetic moment of 6·0 B.M., whereas

$K_3Fe(CN)_6$ with the strongly polarizable CN groups as the ligands has a moment of only 2·3 B.M., supports this view. However, ferric trisacetylacetonate which, from its physical properties, is obviously a covalent compound, also has a moment of 5·9 B.M. Such observations have led to a modification of the original division into ionic and covalent complexes. It is now

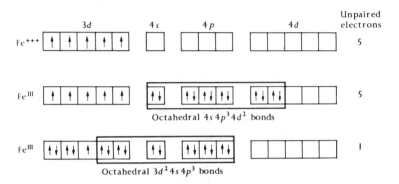

Fig. 2. A comparison of the electronic structure of trivalent iron in 'inner d orbital' and 'outer d orbital' octahedral complexes.

believed that the so-called ionic complexes involve $4d$ rather than $3d$ orbitals, and that the hybrid orbitals should be written $4s4p^34d^2$, whereas the covalent complexes still involve $3d^24s4p^3$ octahedral orbitals. This has led to a classification of these complexes into 'outer d orbital type' and 'inner d orbital type'. It can be seen from Fig. 2 that these ideas will explain, in this case, the observed values of the magnetic moments. Our knowledge of the type of bonding in metallic complexes is by no means complete and much work is being done in this field. Nevertheless, there is no doubt that measurements of paramagnetism can distinguish two types of bond.

Paramagnetism can, in certain cases, give useful information about the stereochemical arrangement of the ligands around the central metal atom. Divalent nickel, with two unpaired electrons in the free ion, can form two types of complex with four ligands. If the bonds are arranged tetrahedrally and involve sp^3 hybridized orbitals, the complex should be paramagnetic with a magnetic moment corresponding to two unpaired electrons. If, however, square-planar bonds are formed with dsp^2 orbitals, the complex should be diamagnetic. The dispositions of the electrons are shown in Fig. 3. This striking difference in magnetic properties has been used to show that the arrangement around the nickel atom in $Ni(NH_3)_4SO_4$, $Ni(N_2H_4)_2SO_4$, and nickel

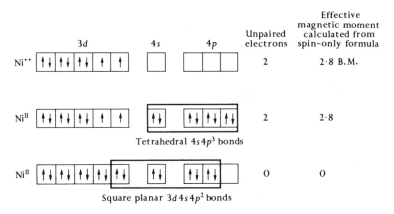

FIG. 3. A comparison of the electronic structure of divalent nickel in tetrahedral and square-planar complexes.

bisacetylacetonate, for example, is different from the planar arrangement found in $K_2Ni(CN)_4$, nickel glyoxime, nickel phthalocyanine, and many other compounds.

In a similar manner divalent cobalt with three unpaired electrons can form tetrahedral complexes in which no electron pairing takes place, or square-planar complexes with only one unpaired electron as shown in Fig. 4. Thus a magnetic moment of about 3·9 B.M. indicates a tetrahedral complex as in $Co(N_2H_4)_2Cl_2$, whilst a moment of about 1·7 B.M. indicates a planar complex as in cobaltous phthalocyanine. In fact the moments observed for the cobaltous complexes are rather higher than the values calculated from the spin-only formula, but this is due to an appreciable orbital contribution. In spite of this additional complication the two types of complex can easily be distinguished.

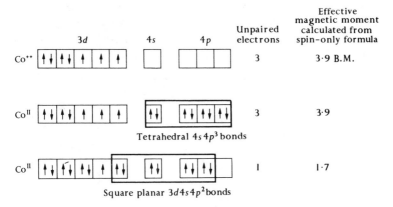

FIG. 4. A comparison of the electronic structure of divalent cobalt in tetrahedral and square-planar complexes.

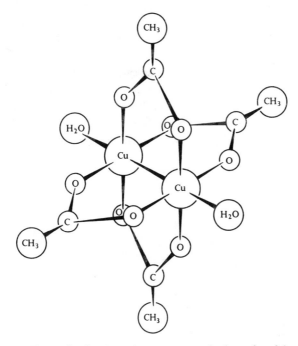

FIG. 5. The molecule of cupric acetate monohydrate found in the solid state.

Magnetic evidence can give unequivocal evidence of the stereochemistry of only a few metal atoms. However, magnetic measurements have proved valuable for indicating the existence of unsuspected bonds in a number of molecules. If cupric acetate monohydrate, $Cu(CH_3COO)_2H_2O$, were like other cupric salts, it would be paramagnetic with a moment corresponding to one unpaired electron. It is, however, diamagnetic. An X-ray structure analysis of this salt has shown that the molecules exist as dimeric units of formula $Cu_2(CH_3COO)_4 2H_2O$ (Fig. 5). The two copper atoms in each unit are only $2 \cdot 64$ Å apart and presumably some sort of electron pairing takes place with the formation of a metal-metal bond.[†] $Fe_2(CO)_9$ is another example of a molecule which contains two atoms which would normally be paramagnetic, and yet the substance exhibits diamagnetism. Again an X-ray analysis of the crystal shows a close approach of the two metal atoms (Fig. 6).[‡] The same sort of interaction has also been shown to occur in $Co_2(CO)_8$, which is also diamagnetic.[§]

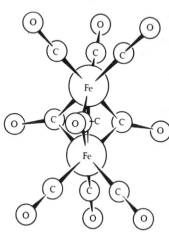

FIG. 6. The molecule of iron enneacarbonyl, $Fe_2(CO)_9$.

REFERENCES

P. W. SELWOOD, *Magnetochemistry* (Interscience).
R. S. NYHOLM, *Q. Rev. chem. Soc.* 1953, **7**, 377.

   † J. N. van Niekerk and F. R. L. Schoening, *Acta crystallogr.* 1953, **6**, 227.
   ‡ H. M. Powell and R. V. G. Ewens, *J. chem. Soc.* 1939, 286.
   § G. G. Gardner, H. P. Klug, and L. E. Alexander, *Acta crystallogr.* 1964, **17**, 732.

XIII

NUCLEAR MAGNETIC RESONANCE

13.1. Nuclear magnetic moments and susceptibilities

THIS chapter could equally well be included in the previous chapter or in the part dealing with spectroscopy, since, as we shall see, nuclear magnetic resonance is concerned with the detection by a spectroscopic method of changes in susceptibility. In the last chapter we discussed the bulk susceptibility which arises largely from the presence of extranuclear electrons in the atoms of the sample molecules. However, in order to account for the hyperfine structure of spectral lines, it is necessary to postulate that many atomic nuclei have an intrinsic angular momentum or spin. This nuclear spin can give rise to nuclear paramagnetism.

A rotating sphere of mass M and charge e has associated with it a magnetic moment $pe/2Mc$, where p is the angular momentum. For nuclei this expression has to be modified, so that

$$\mu_a = gpe/2M_p c, \tag{1}$$

where μ_a is the *actual magnetic moment*, M_p is the mass of the proton, and g is called the *nuclear g factor*. g is a small non-integral number which has to be determined experimentally. It can have either positive or negative values. For example g for a proton is $+5\cdot585$ and for a neutron $-3\cdot826$. A negative sign denotes that the moment behaves like that of a negatively charged particle. We shall call the direction along which the actual magnetic moment μ_a acts the *magnetic axis*. We shall see later that the quantity which is normally listed as the 'magnetic moment' is, for reasons of convenience, not μ_a, but a quantity closely related to μ_a. Equation (1) is empirical, and nuclear theory has not yet advanced sufficiently to provide an explanation for it. However, it does suggest that the spins of most of the particles that constitute the nucleus pair off, leaving only a small number to provide the nuclear spin. This is confirmed by the fact that all nuclei with even atomic number and even mass number, which represent about two-thirds of all stable nuclei, have zero magnetic moment. Such nuclei cannot be studied by the nuclear magnetic resonance method, which depends on the presence of a nuclear moment.

The quantum theory requires the angular momentum of the nucleus to have only discrete values given by

$$p = (h/2\pi)\sqrt{\{I(I+1)\}}, \qquad (2)$$

where I is the spin of the nucleus, expressed in units of $h/2\pi$. I is an integer for nuclei with even atomic number and a half-integer for odd atomic numbers. From equations (1) and (2) we may eliminate p and obtain

$$\mu_a = \frac{geh}{4\pi M_p c}\sqrt{\{I(I+1)\}} = g\mu_n\sqrt{\{I(I+1)\}}. \qquad (3)$$

μ_n is the unit of nuclear magnetic moment and is called the *nuclear magneton*. Since

$$\mu_n = \frac{eh}{4\pi M_p c} = 5\cdot05038 \times 10^{-24} \text{ erg/gauss}$$

and M_p is about 2000 times as great as the mass of an electron, it can be seen from equation (12.5) that nuclear magnetic moments will be less than electronic magnetic moments by a factor of about 10^3.

The nuclear paramagnetic susceptibility can be represented by an expression similar to equation (12.2) for electronic paramagnetic susceptibility,

$$\chi_N = N_0 g^2 \mu_n^2 I(I+1)/3kT.$$

Since this equation involves the square of the nuclear magneton, nuclear susceptibilities will be smaller than electronic susceptibilities by a factor of about 10^6.

13.2. Nuclear magnets in external magnetic fields

From one point of view we may regard the nucleus as a bar magnet with an actual magnetic moment given by equation (3). When a nucleus is placed in a uniform external magnetic field, it will, just like any other magnet, suffer a torque which will tend to align the magnetic axis along the direction of the external field. However, since the nuclear magnet possesses spin angular momentum, we must also associate with the nucleus the properties of a gyroscope in a gravitational field. In the same way that a spinning top precesses under the torque of a gravitational field, the nucleus will precess under the torque of the magnetic field. The frequency ν_0 with which the magnetic axis of the nucleus precesses about the external field direction is known as the *Larmor frequency*.

The potential energy of the nuclear magnet of moment μ_a in an external field H_0 is

$$V = -\mu_a H_0 \cos\theta = -\mu_H H_0, \qquad (4)$$

where θ is the angle between μ_a and H_0, and μ_H is the component of μ_a in the direction of the external field H_0. The quantum theory requires this energy to adopt only certain discrete values, and the condition is that the components of the angular momentum in the direction of the external field shall be $mh/2\pi$, where $m = I, I-1,..., +1, 0, -1,..., -I$. Thus there are $2I+1$ values of m and hence $2I+1$ orientations of the nuclear magnet with respect to the direction of the applied field.

Since the actual magnetic moment is given by equation (1), and since the component of the angular momentum p can have only the values $mh/2\pi$, the component μ_H of the actual magnetic moment along the field is restricted to the values

$$\mu_H = \frac{ge}{2M_p c} \frac{mh}{2\pi}$$

$$= gm \frac{eh}{4\pi M_p c}$$

$$= gm\mu_n. \qquad (5)$$

The maximum component of μ_a along the field occurs when m has its maximum value I. Hence

$$\mu_H(\text{max}) = gI\mu_n = \mu\mu_n.$$

The quantity μ is what is usually called the 'magnetic moment' of the nucleus. In Table XLIX some of the lighter nuclei are listed together with their spin I in units of $h/2\pi$, their g factor in units of the nuclear magneton $eh/4\pi M_p c$, and the quantity gI which is usually called the 'magnetic moment'.

TABLE XLIX

The spin, g factor, and magnetic moment of some of the light nuclei

Nucleus	g factor	Spin (I)	Magnetic moment $\mu = gI$
^1n	-3.826	$\frac{1}{2}$	-1.913
^{1}H	$+5.585$	$\frac{1}{2}$	$+2.793$
^{4}He	0	0	0
^{7}Li	$+2.171$	$\frac{3}{2}$	$+3.257$
^{9}Be	-0.785	$\frac{3}{2}$	-1.178
^{10}B	$+0.600$	3	$+1.801$
^{12}C	0	0	0
^{13}C	$+1.405$	$\frac{1}{2}$	$+0.702$
^{14}N	$+0.404$	1	$+0.404$
^{15}N	-0.566	$\frac{1}{2}$	-0.283
^{16}O	0	0	0
^{19}F	$+5.256$	$\frac{1}{2}$	$+2.628$

13.3. Nuclear magnetic resonance

For a nucleus of spin $\frac{1}{2}$, such as ^{1}H or ^{19}F, there will be two possible orientations of the nuclear magnet, as shown in Fig. 1. This figure shows one state of higher energy in which the component of the actual magnetic moment in the direction of the field is $-\mu_H$ and the other of lower energy in which the component is $+\mu_H$. Since m can have only the two values $+\frac{1}{2}$ or $-\frac{1}{2}$, then, from equation (5),

$$\mu_H = \pm \tfrac{1}{2} g\mu_n.$$

From equation (4), the energy difference between the two states is

$$\Delta V = g\mu_n H_0, \qquad (6)$$

which is the work required to reverse the direction of a magnet of strength $g\mu_n/2$ which had initially been aligned along the direction of the field H_0. If transitions occur between the two states, the frequency of the absorbed or emitted radiation will be $g\mu_n H_0/h$, and this frequency turns out to be identical with the Larmor precession frequency ν_0. The two energy levels are very closely spaced and, if thermal equilibrium exists, the population difference between them is only three or four nuclei in every million. This minute population difference means that only very faint signals could be obtained by conventional absorption or emission spectroscopy, signals which would be far below the

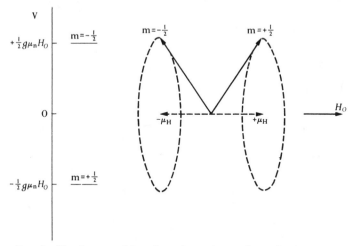

FIG. 1. The two possible orientations of a nucleus of spin $\frac{1}{2}$ in a magnetic field.

limit of detection. For example, if a beam of radiation of the correct frequency is passed into a sample of ice placed in a magnetic field of strength about 7000 gauss, it will have to traverse about 100 km before its intensity is reduced to half the original value.

In order to get round this difficulty we may take advantage of the resonance effect. It is well known that some large structures, such as bridges, which have stood up to very high winds have been destroyed by quite gentle gusts if these gusts happen to strike with the natural frequency of the structure. It is for this reason that troops of marching soldiers are told to break step when passing over footbridges. In the present case, instead of avoiding the resonance condition, we deliberately try to achieve it. We irradiate our sample with electromagnetic waves of the Larmor frequency and we can observe effects out of all proportion to those observable with a slightly different frequency. In fact what happens is that, at the resonance frequency, the

nuclear magnets will topple very readily from one to the other
of the states shown in Fig. 1. Consequently, we obtain a sort of
stimulated absorption which is of a detectable magnitude.

We may also take advantage of the fact that, according to
equation (6), the energy difference between the two magnetic
states of our nucleus depends on the strength of the external
magnetic field. If we again consider hydrogen nuclei with
$g = 5 \cdot 585$, and substitute in the expression

$$\nu_0 = g\mu_n H_0/h,$$

remembering that $\mu_n = eh/4\pi M_p c$, we find for an external field
of 10,000 gauss

$$\nu_0 = \frac{5 \cdot 585 \times 4 \cdot 803 \times 10^{-10} \times 10,000 \times 6 \cdot 025 \times 10^{23}}{4 \times 3 \cdot 142 \times 1 \cdot 0076 \times 3 \times 10^{10}} \text{ c/s}$$

$$= 42 \cdot 6 \times 10^6 \text{ c/s}.$$

In a field of 10,000 gauss, therefore, the resonance frequency
will be $42 \cdot 6$ Mc/s. Similarly, the resonance frequency will be
30 Mc/s in a field of 7050 gauss and 10 Mc/s in a field of 2350
gauss. Consequently, we can perform a nuclear magnetic
resonance experiment in two ways. We can apply a static
external magnetic field H_0, and an oscillatory magnetic field H_1
at right angles to the main field H_0. The periodicity of the field
H_1 is varied until it becomes identical with the Larmor frequency
in the static field H_0. At that frequency, and only at that frequency, will detectable absorption by the sample nuclei occur.
Alternatively we can apply an oscillatory magnetic field with a
fixed periodicity, and vary the external field H_0 until again the
state of resonance is reached. In either case we measure the
change in nuclear paramagnetic susceptibility that occurs at
the condition of resonance. The spectrum may be plotted as
intensity of absorption against either frequency or applied field
strength, depending upon which experimental method is used.
In most nuclear magnetic resonance spectrometers a static field
H_0 is applied by means of a large permanent magnet or electro-
magnet. This field can be increased or decreased by about 20
gauss when a current is passed through subsidiary coils. The
actual external field of strength H thus varies slightly, but is

always close to the static value H_0. The majority of experimental results are plotted not against frequency but against ΔH, where ΔH is the variation in the applied field strength and is given by
$$\Delta H = H - H_0.$$
We shall always present results in this way.

Even when this resonance method is used, the signals are still very weak compared with more conventional spectroscopic signals. Great care must be taken to ensure a homogeneous magnetic field, and to exclude extraneous magnetic effects. Furthermore, the electronic circuits which are used for supplying power and for detecting the change in nuclear paramagnetic susceptibility must be designed and maintained with great thoroughness.

13.4. The line shape

For isolated nuclei the absorption spectrum will be a single line of extreme sharpness as shown in Fig. 2 (a). In some nuclear resonance spectrometers it is found to be more convenient to plot the first derivative of the absorption curve with respect to applied field strength or frequency. In such cases a curve of the type shown in Fig. 2 (b) is obtained. The true absorption curve may readily be obtained from the derivative curve by graphical integration. For nuclei isolated in pairs, the interaction of one nucleus with the other will modify the spectrum. If we consider two identical nuclei of spin $\tfrac{1}{2}$ and magnetic moment μ at a distance r apart in an external magnetic field H_0, then the effective field at one nucleus is given by

$$H_{\text{eff}} = H_0 \pm \frac{3\mu\mu_{\text{n}}}{2r^3}(3\cos^2\theta - 1),$$

where θ is the angle between the direction of H_0 and the line joining the two nuclei. Consequently the absorption spectrum will be a doublet with a separation of

$$\Delta H = \frac{3\mu\mu_{\text{n}}}{r^3}(3\cos^2\theta - 1). \tag{7}$$

In a finely divided powdered solid the absorption line is an

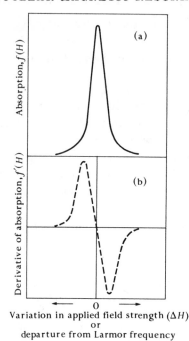

FIG. 2. Resonance curves for isolated nuclei
(a) absorption curve, (b) derivative curve.

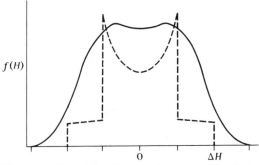

FIG. 3. The absorption curve of a two-spin system. The dotted curve
gives the absorption for an isolated pair of nuclei. The full curve
shows the effect of neighbouring nuclei on the isolated system.
(Adapted from G. E. Pake, *J. chem. Phys.* 1948, **16**, 327.)

average resulting from the superposition of the lines from all
possible values of the angle θ. In reality the ideal shape, which
is shown as a dotted line in Fig. 3, is broadened by the effect of

the more distant nuclei. The broadening may be represented by a factor of the form

$$\exp\{-(\Delta H)^2/2\beta^2\},$$

where β^2 is an arbitrary constant which may be adjusted so as to allow for the effects of intermolecular broadening. The continuous line in Fig. 3 shows the effect of the broadening term on

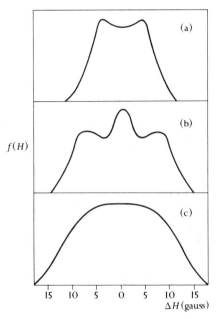

FIG. 4. Comparative absorption curves: (a) two-spin system, (b) three-spin system, (c) four-spin system.

the dotted curve which allowed only for nearest-neighbour interaction. Similar curves have been calculated for an equilateral triangle and for a regular tetrahedron of nuclei of spin $\frac{1}{2}$. The results of these calculations for the so-called two-, three-, and four-spin systems are shown graphically in Fig. 4.[†] In each case the same value for r has been assumed and the three drawings are all on the same scale. Changes in r will alter the

† C. M. Deeley and R. E. Richards, *J. chem. Soc.* 1954, 3697.

width of the curves, but the same general contours will be maintained.

An example of the use of such curves to solve a problem in molecular structure is provided by a nuclear resonance study of the solid that is precipitated when ammonia is added to a solution of mercuric chloride under controlled conditions.† The deposit is

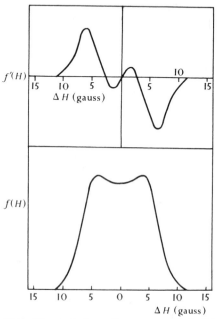

FIG. 5. Derivative and absorption curves of infusible white precipitate, NH_2HgCl.

called 'infusible white precipitate' and its structure has been in doubt for many years. The three most probable suggestions are NH_2HgCl, $xHgO.(1-x)HgCl_2.2NH_3$, and $NHg_2Cl.NH_4Cl$. It will be seen that these three formulae contain groups of two, three, and four protons respectively. The derivative and the reconstituted absorption curves of infusible white precipitate are shown in Fig. 5, and they are clearly of the form expected for a two-proton system. The proton resonance spectrum thus supports the formula NH_2HgCl. A similar type of application

† C. M. Deeley and R. E. Richards, *J. chem. Soc.* 1954, 3697.

is found in the study of the monohydrate of nitric acid.† The proton resonance absorption spectrum is shown in Fig. 6, and the shape of the curve is that expected for a three-proton system. The correct formula of the monohydrate is, therefore, $H_3O^+NO_3^-$ and not $HNO_3.H_2O$.

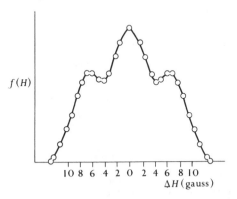

FIG. 6. Absorption curve of nitric acid monohydrate, $H_3O^+NO_3^-$.

13.5. Second moments and the determination of molecular parameters

Information about the actual distance between the resonating nuclei may be obtained by the use of an equation for the mean-square width of the absorption line, which is a measure of the broadening of the line. The mean square width, or *second moment* as it is usually called, is given by

$$\langle \Delta H^2 \rangle_{av} = \tfrac{3}{2}I(I+1)g^2\mu_n^2 N^{-1} \sum_{j>k} (3\cos^2\theta_{jk}-1)^2 r_{jk}^{-6} +$$
$$+ \tfrac{1}{3}\mu_n^2 N^{-1} \sum_j \sum_f I_f(I_f+1)g_f^2(3\cos^2\theta_{jf}-1)^2 r_{jf}^{-6}. \quad (8)$$

In this equation the f subscripts refer to nuclei (with a magnetic moment) of a different kind to those at resonance, N is the number of nuclei at resonance in the unit cell of the crystal, θ_{jk} is the angle between the line joining the jth and kth resonating nuclei and the direction of the applied magnetic field, and r_{jk} is the distance between these two nuclei. For a powdered sample

† R. E. Richards and J. A. S. Smith, *Trans. Faraday Soc.* 1951, **47**, 1261.

this equation may be simplified slightly by replacement of $(3\cos^2\theta - 1)^2$ by its spatial average of $\tfrac{4}{5}$. We obtain

$$\langle \Delta H^2 \rangle_{\mathrm{av}} = \tfrac{6}{5} I(I+1) g^2 \mu_{\mathrm{n}}^2 N^{-1} \sum_{j>k} r_{jk}^{-6} +$$
$$+ \tfrac{4}{15} \mu_{\mathrm{n}}^2 N^{-1} \sum_{j} \sum_{f} I_f(I_f+1) g_f^2 r_{jf}^{-6}. \qquad (9)$$

The second moment $\langle \Delta H^2 \rangle_{\mathrm{av}}$ may readily be obtained from the experimental absorption curve through the relation

$$\langle \Delta H^2 \rangle_{\mathrm{av}} = \int_0^\infty f(H) \cdot \Delta H^2 \, dH \Big/ \int_0^\infty f(H) \, dH, \qquad (10)$$

where $f(H)$ is the intensity of the absorption line as a function of H, and ΔH is, as before, the deviation from the centre of the symmetrical resonance line. Equation (10) merely means that, to get the second moment from the absorption curve, we find the moment of inertia of this curve about its symmetry axis and divide by the area under the absorption curve. A convenient mechanical device known as an integrator, which will determine second moments directly from the absorption or derivative curves, is normally used in nuclear magnetic resonance work.

Before we consider an example of the determination of internuclear distances, we must consider equations (8) and (9) in rather more detail. The second moment depends on the inverse sixth power of the internuclear distance. Consequently, an error of 6 per cent in the second moment involves an error of only 1 per cent in r_{jk}. Nuclear magnetic resonance methods should, therefore, yield accurate values for internuclear distances. However, there are some drawbacks to the method. We have already seen that about two-thirds of the total number of stable nuclei have no magnetic moment and so cannot be studied by this method. Practically every element does, in fact, have a stable isotope with a nuclear magnetic moment, but sometimes these isotopes have a very low percentage abundance. Equations (8) and (9) are useful only for nuclei with spin $\tfrac{1}{2}$, since nuclei with spin of 1 or more possess nuclear quadrupole moments which bring about a further splitting of the absorption spectrum and thus affect the second moment. These equations are valid only if the contents of the crystalline unit cell are rigid at the temperature at which the experiment is performed, since internal motions such as free rotation narrow the absorption line. They are accurately valid only if the number of paramagnetic impurity atoms is less than about one in every million resonating nuclei, since the oscillatory magnetic fields produced by the thermal agitation of the paramagnetic atoms will broaden the absorption line. Finally, apart from the resonating nuclei, it is almost essential to know the positions in the unit cell of all other atoms with magnetic moments, particularly if these moments happen to be large. If the moments are small it is sometimes possible to guess reasonable values of the broadening function β^2, but it is always a great help, and it certainly adds to the reliability of the results, if the various r_{jf}'s are known. These restrictions are rather severe, and the determination of molecular parameters by nuclear magnetic-resonance methods has been confined almost exclusively to those involving hydrogen atoms, since hydrogen atoms are neither readily nor accurately located by X-ray diffraction.

13.6. The determination of molecular parameters in powdered samples

As a typical example of the use of nuclear magnetic resonance to find molecular parameters in powdered samples, we may consider the determination of the P–H distance in phosphonium iodide, PH_4I.† The proton-resonance spectrum of the powdered specimen was recorded as a derivative curve at 90° K, and is shown in Fig. 7. From this curve the observed second moment was found to be 14·56 gauss². If the phosphonium ion is assumed to be tetrahedral, use may be made of equation (9) to obtain calculated values of the second moment. Since all three types of atoms in phosphonium iodide have magnetic moments, allowance must be made, through the last term in equation (9), for broadening due to I...H and P...H interaction, as well

† L. Pratt and R. E. Richards, *Trans. Faraday Soc.* 1954, **50**, 670.

as for the main H...H interaction. A small correction has to be made for the contraction of the lattice which occurs when the crystal is cooled to 90° K. An attempt was also made to correct $r_0(PH)$ to $r_e(PH)$ by allowance for the zero-point motion of the atoms. Since there is some doubt as to the usefulness and the validity of this correction, we shall not discuss it further.

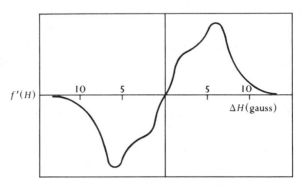

FIG. 7. The derivative curve of phosphonium iodide, PH_4I.

Nevertheless, since the correction is small and is without doubt of the right order of magnitude, we shall include it with the other corrections and quote the final results as r_e values. Finally, the interionic contributions depend to some extent on the orienta-

TABLE L

Observed and calculated second moments of phosphonium iodide

$r_e(PH)$	1·40 Å	1·41 Å	1·42 Å	1·43 Å
Intra-ionic H . . . H broadening .	7·521	7·206	6·907	6·622
Vibration correction . . .	0·417	0·394	0·372	0·352
Intra-ionic P . . . H broadening.	3·469	3·324	3·184	3·055
Vibration correction . . .	0·148	0·140	0·132	0·124
Interionic H . . . H broadening .	2·969	3·000	3·032	3·064
Interionic I . . . H broadening .	0·598	0·608	0·618	0·628
Interionic P . . . H broadening .	0·033	0·033	0·034	0·034
Lattice contraction	0·252	0·255	0·257	0·261
Total second moment . . .	15·407	14·960	14·536	14·140
Observed second moment . .	14·56	14·56	14·56	14·56

All broadenings and second moments are in gauss².

tion of the phosphonium ions in the unit cell. In Table L the various factors contributing to the broadening of the absorption line are shown. The phosphonium ion was assumed to lie with the P—H bonds directed towards the nearest iodine ions, which are approximately tetrahedrally disposed about the phosphorus atom. Other possible orientations of the tetrahedral phosphonium ion lead to second moments which differ by not more than $\pm 0 \cdot 3$ gauss^2 from the values in Table L. From the calculated second moments it may be concluded that $r_e(\text{PH}) = 1 \cdot 42 \pm 0 \cdot 02$ Å.

It should be noticed that the distance obtained directly from nuclear magnetic resonance experiments is seldom an actual bond length. It is usually a distance between two atoms both of which are bonded to a third atom but not to each other. In the studies of the proton resonance of hydrates, for instance, it is the interproton distance that is determined and $r_0(\text{OH})$ cannot be found without a knowledge of the HOH angle. In some compounds, such as hydrazine fluoride, $N_2H_6F_2$, it is possible to observe the resonance spectrum of both the hydrogen and the fluorine nuclei, and hence to determine two unknown parameters.† In this way it was found that the H...F distance was $1 \cdot 542$ Å and $r_0(\text{NH})$ was $1 \cdot 075$ Å on the assumption that the hydrogen atoms lie along the N–H...F hydrogen bonds. That the hydrogen atoms do lie along this line is confirmed by the fact that the N...F distance determined by means of X-rays is $2 \cdot 62$ Å,‡ which is equal, within experimental error, to the sum of the H...F distance and $r_0(\text{NH})$ determined by nuclear magnetic resonance. A list of some of the crystalline powders studied quantitatively by nuclear magnetic resonance is given in Table LI. In some of these compounds it would be very difficult to locate the hydrogen atoms by means of X-rays owing to the large scattering of the metal atoms. However, a neutron diffraction investigation (Chapter IX) would give rather more information.

† C. M. Deeley and R. E. Richards, *Trans. Faraday Soc.* 1954, **50**, 560.
‡ M. L. Kronberg and D. Harker, *J. chem. Phys.* 1942, **10**, 309.

TABLE LI

Selection of interproton distances found by nuclear magnetic resonance

Compound	H ... H *distance*
$Li_2SO_4 . H_2O$	1·59 Å
$CuCl_2 . 2H_2O$	1·60
$(NH_4)_2CuCl_4 . 2H_2O$	1·59
$K_2CuCl_4 . 2H_2O$	1·62
$CaSO_4 . 2H_2O$	1·58
$Ba(ClO_3)_2 . 2H_2O$	1·56
$K_2HgCl_4 . H_2O$	1·607
$K_2SnCl_4 . H_2O$	1·62
$(COOH)_2 . 2H_2O$	1·65
C_6H_6	2·495

13.7. The use of single crystals

More information may be gained from the nuclear magnetic resonance spectrum if a single crystal is used instead of a powder, and spectra are recorded with the crystal in different known orientations with respect to the external magnetic field. If such measurements are made with a single crystal of a simple hydrate such as $CaSO_4 . 2H_2O$ it is possible to obtain the direction in the unit cell of the line joining the two protons in each water molecule as well as the interproton distance.† A number of experiments of this sort have been carried out, and we may illustrate the method by a consideration of urea, $(NH_2)_2CO$.‡ We discussed in section 7.5 how X-ray analysis has shown that urea crystallizes in a tetragonal unit cell with $a = b = 5·66$ Å and $c = 4·41$ Å. All the C—O bonds are aligned along the four-fold axis $[c]$. From the space group the molecule must have mm (C_{2v}) symmetry. All the atoms in the urea molecule have now been located both by X-ray and neutron diffraction, and the molecule has been shown to be planar. Nevertheless the problem of distinguishing the two possible configurations of mm (C_{2v}) symmetry, the non-planar form (Fig. 7.10) and the planar form

† G. E. Pake, *J. chem. Phys.* 1948, **16**, 327.
‡ E. R. Andrew and D. Hyndman, *Discuss. Faraday Soc.* 1955, **19**, 195.

(Fig. 8), has all the ingredients necessary for a successful solution by means of nuclear magnetic resonance, and we shall use it to illustrate the application of single crystals.

Two single crystals weighing rather over 1 g were grown and mounted in different orientations. In the first the fourfold axis $[c]$ was made the vertical axis of rotation and was set perpendicular to the external magnetic field H_0. In the second the diagonal of the basal plane $[1\bar{1}0]$ was used as the axis of rotation. Derivative curves were then obtained for various azimuthal angles ϕ between the field direction and $[b]$ with the first crystal, and for ψ between the field direction and $[c]$ with the second. The relations between the angles and the axes are shown in Fig. 9. When $\phi = 0$ the external field is along $[b]$ and when $\phi = 45°$ the field is along $[1\bar{1}0]$; when $\psi = 0$ the field is along $[c]$ and when $\psi = 90°$ the field is along $[110]$ which is at right angles to the vertical $[1\bar{1}0]$. The experimental values of the second moment as a function of ϕ or ψ are shown in Fig. 10. In each part of the figure calculated curves A and B are given. A refers to the non-planar model and B to the planar model. It is clear that only the planar model reproduces the experimental values satisfactorily.

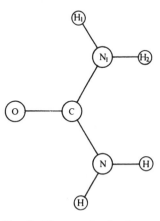

FIG. 8. Planar molecule of urea, $(NH_2)_2CO$.

The next step in the analysis was to determine the positions of the protons from the experimental values of the second moments. Four parameters are required to fix these positions: $r_0(N_1H_1)$, $r_0(N_1H_2)$, and the angles $H_1N_1H_2$ and CN_1H_1. Since only three independent experimental values of the second moment are available it was assumed that $r_0(N_1H_1) = r_0(N_1H_2)$. Calculated second moments were obtained from equation (8) with various values of the three parameters. The calculated moments are compared with the experimental moments in Table LII. The

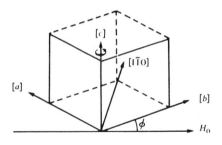

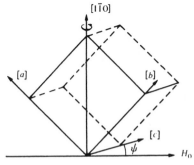

FIG. 9. The axes used for the description of the setting of urea crystals.

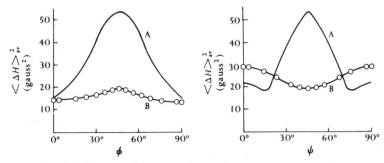

FIG. 10. The second moment of urea as a function of orientation.

values of the parameters which best reproduce the observed second moments are $r_0(NH) = 1 \cdot 046$ Å, $H_1N_1H_2 = 119 \cdot 1°$, and $CN_1H_1 = 120 \cdot 5°$. The bond length is estimated to be accurate to $\pm 0 \cdot 01$ Å and the angles to $\pm 2°$. The H...H distance is $1 \cdot 803 \pm 0 \cdot 015$ Å.

TABLE LII

Calculated second moments of urea

r_0(NH)	$H_1N_1H_2$	CN_1H_1	Values of second moments in gauss^2		
			field along [b]	field along [110]	field along [c]
1·040 Å	120°	120°	13·5	18·6	29·3
1·020	120°	120°	13·9	19·6	31·8
1·040	122°	120°	12·9	19·2	27·2
1·040	120°	118°	13·8	17·8	31·8
1·046	119·1°	120·5°	13·6	18·2	29·0
Mean experimental values			13·6	18·2	29·0

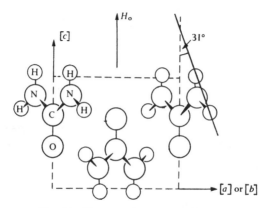

FIG. 11. The crystal structure of urea.

As a final check on the structure, the separation of the doublet of the absorption curve when the external field is parallel to [c] was calculated from equation (7). The space group requires the planes of alternate molecules to be at right angles to each other and, if the planar model is correct, all the lines joining neighbouring pairs of protons make the same angle $\theta = 31°$ with [c] as shown in Fig. 11. On the other hand, if the non-planar model were correct, θ would be 90°. The calculated separation for the planar model is 8·7 gauss and for the non-planar model 7·2 gauss, for an interproton distance of 1·803 Å. The experimental value of the separation when the external field is parallel to [c] is 8·9 gauss, which again confirms the planar model.

13.8. Internal motion in solids

Nuclear magnetic resonance may be used to detect and interpret internal motion in solids. In the determination of interproton distances, experiments are usually performed at low
temperatures to ensure the absence of large thermal motions. If
the temperature at which the experiment is performed is raised,
a sudden drop in the value of the second moment is occasionally
observed. The narrowing of the absorption line is usually a result
of the onset of ionic or molecular reorientation. Thus in the four-
spin system potassium borohydride, KBH_4,† at 20° K the second
moment of the absorption curve is 36·23 gauss^2 and the line
width, defined as the separation of the two maxima of the
derivative curve, is 15·5 gauss. These figures are maintained up
to about 85° K when they suddenly fall to 3·05 gauss^2 and
4·5 gauss respectively. It is clear that some sort of change occurs
in the structure of the crystal at 85° K. The narrowing of the
line in KBH_4 is found at a temperature not far removed from the
λ-point in the specific heat curve. However, this is apparently
fortuitous, since in $NaBH_4$ the line-width transition is over

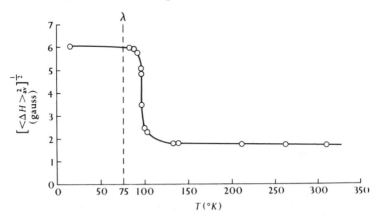

Fig. 12. Plot of the square root of the second moment of potassium
borohydride, KBH_4, against the absolute temperature.

† T. Ford and R. E. Richards, *Discuss. Faraday Soc.* 1955, **19**, 230.

$100°$ C below the λ-point. It is evident that the different experimental techniques are detecting different sorts of reorientation, and it is probable that nuclear magnetic resonance is detecting small jumps of the BH_4^- ions about random axes, rather than the onset of free rotation. The experimental plot of the square root of the second moment against temperature shown in Fig. 12 may be interpreted in terms of a barrier height of $3\cdot76$ kcal/mole.

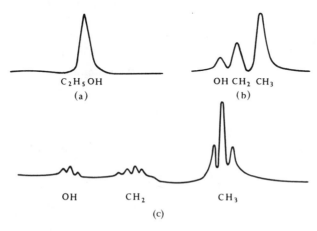

FIG. 13. The proton resonance spectrum of ethyl alcohol showing the chemical shift: (a) under low resolution, (b) under medium resolution, (c) under high resolution.

13.9. High resolution nuclear magnetic resonance

So far we have dealt only with broad line nuclear magnetic resonance spectra obtained from solids because such studies have yielded values of molecular parameters. In fact the application of nuclear magnetic resonance to liquids and solutions is of vastly greater importance in chemistry. In the less condensed states of matter it is found that the relatively free molecular motion produces very narrow absorption lines and that the resonance frequency for a given molecular species is not constant but depends on the compound in which the nucleus is present. Moreover, identical nuclei in the same molecule may resonate at slightly different frequencies. This so-called *chemical shift* is

illustrated in Fig. 13.† Fig. 13 (a) shows the proton resonance
absorption spectrum of ethyl alcohol under low resolution.
When the resolution is increased, as in Fig. 13 (b), the line splits
into three components. The areas under the curves are in the
ratio of 3 : 2 : 1, and the components are assigned to the groups
CH_3, CH_2, and OH respectively. Under very high resolution,
as in Fig. 13 (c), each component shows a multiplet structure
which is consistent with the previous assignment. In order to
account for the fine structure we have to remember that the
magnetic field at the nucleus may not be equal to the external
field, but may be modified slightly by the presence of molecular
electrons. These electrons will produce a small field in addition
to the applied one. If H_N is the field at the nucleus we may write

$$H_N = H_0 - \sigma H_0, \qquad (11)$$

where σ is the *shielding constant* and defines the chemical shift.
For protons σ is about one-millionth of H_0 and may be either
positive or negative. It will be different for every resonating
nucleus in a different electronic environment.

Chemists concerned with the preparation of new compounds,
particularly organic compounds, have found chemical shifts
invaluable. By a survey of related molecules they have been able
to correlate the chemical shifts of resonating nuclei with known
structural features and thus have a tool which can be used to
predict how the atoms in an unknown compound are linked
together. Much of the work has been concerned with the study
of protons, but it is rapidly being extended to other elements
such as boron, carbon, nitrogen, fluorine, silicon, and phosphorus.
Such an empirical approach can give extensive information about
extremely complex organic molecules.

In the realm of stereochemistry the same approach has been
used, for example, to distinguish *cis* and *trans* geometrical isomers
in compounds containing a double bond. Of more interest to us
is the possibility of determining molecular symmetry without
recourse to a knowledge of the magnitude of chemical shifts.

† J. E. Wertz, *Chem. Rev.* 1955, **55**, 829.

As an illustration we may consider the three dichlorobenzenes (Fig. 14). *Ortho*-dichlorobenzene (Fig. 14 (a)) has two sets of protons in different environments and gives a spectrum known

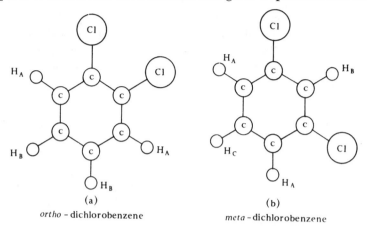

(a)
ortho - dichlorobenzene

(b)
meta - dichlorobenzene

(c)
para - dichlorobenzene

FIG. 14. The three isomeric dichlorobenzenes, showing the different proton environments.

as an A_2B_2 type. *Meta*-dichlorobenzene (Fig. 14 (b)) has protons in three non-equivalent sites and gives an A_2BC spectrum. Finally *para*-dichlorobenzene (Fig. 14 (c)) has all four protons equivalent and gives a single peak. Thus it is possible to distinguish *ortho-*, *meta-*, and *para*-dichlorobenzene merely by an inspection of their proton resonance spectra.

If the two substituents are different, it can readily be seen that the *para*-substituted compound will give an A_2B_2 spectrum, whereas the *ortho-* and *meta*-substituted compounds will both give $ABCD$ spectra which are too complicated to be differentiated by inspection.

Another source of information which is being actively explored is the distance between the multiplets in each component in, for example, Fig. 13 (c). The separation of the multiplets depends on the extent to which one nucleus interacts magnetically with another, that is on the *coupling constant J*. Coupling constants can be used in much the same way as chemical shifts to obtain information about the way in which atoms in a molecule are linked together, but there are strong prospects that a more fundamental understanding of the magnitudes of coupling constants may be derived from wave mechanics, and that relations between molecular geometry and coupling constants will emerge.

Chemical shifts and coupling constants provide a wealth of experimental information that should be capable of interpretation in terms of the detailed electronic structure of the molecule. The problem is a difficult one but is full of promise, especially as the method is so readily applicable to the liquid state, a state not easily investigated by other experimental methods.

REFERENCES

J. E. WERTZ, *Chem. Rev.* 1955, **55**, 829.

R. E. RICHARDS, *Q. Rev. chem. Soc.* 1956, **10**, 480.

L. M. JACKMAN, *Applications of Nuclear Resonance Spectroscopy in Organic Chemistry* (Pergamon Press).

J. D. ROBERTS, *Nuclear Resonance: Applications to Organic Chemistry* (McGraw-Hill).

J. A. POPLE, W. G. SCHNEIDER, and H. J. BERNSTEIN, *High-resolution Nuclear Magnetic Resonance* (McGraw-Hill).

APPENDIX

Physical constants†

Velocity of light	c	$299793 \cdot 0 \pm 0 \cdot 3$ km s^{-1}
Electronic charge	e	$(4 \cdot 80286 \pm 0 \cdot 00009) \times 10^{-10}$ e.s.u.
Planck's constant	h	$(6 \cdot 62517 \pm 0 \cdot 00023) \times 10^{-27}$ erg s
Boltzmann's constant	k	$(1 \cdot 38044 \pm 0 \cdot 00007) \times 10^{-16}$ erg deg^{-1}
Gas constant‡	R	$(8 \cdot 31696 \pm 0 \cdot 00034) \times 10^{7}$ erg deg^{-1} mole^{-1}
Avogadro's number‡	N_0	$(6 \cdot 02486 \pm 0 \cdot 00016) \times 10^{23}$ mole^{-1}
Standard volume of an ideal gas‡	V_0	$22420 \cdot 7 \pm 0 \cdot 6$ cm^3 mole^{-1}
Atomic mass of the proton‡	M_{p}	$1 \cdot 007593 \pm 0 \cdot 000003$
Atomic mass of the neutron‡	M_{n}	$1 \cdot 008982 \pm 0 \cdot 000003$
Atomic mass of the electron‡	m_{e}	$(5 \cdot 48763 \pm 0 \cdot 00006) \times 10^{-4}$

† From E. R. Cohen, J. W. M. DuMond, and J. S. Rollett, *Rev. Mod. Phys.* 1955, **27**, 363.

‡ On the physical scale.

FORMULA INDEX

SUBJECT INDEX